Energy and Water Management in Western Irrigated Agriculture

Studies in Water Policy and Management
Charles W. Howe, General Editor

Economic Benefits of Improved Water Quality: Public Perceptions of Option and Preservation Values, Douglas A. Greenley, Richard G. Walsh, and Robert A. Young

Municipal Water Demand: Statistical and Management Issues, C. Vaughan Jones, John J. Boland, James E. Crews, C. Frederick DeKay, and John R. Norris

Water and Agriculture in the Western U.S.: Conservation, Reallocation, and Markets, Gary Weatherford, in association with Lee Brown, Helen Ingram, and Dean Mann, eds.

Water and Western Energy: Impacts, Issues, and Choices, Steven C. Ballard, Michael D. Devine, and Associates

Natural Radioactivity in Water Supplies, Jack K. Horner

Flood Plain Land Use Management: A National Assessment, Raymond H. Burby, Steven P. French, Beverly A. Cigler, Edward J. Kaiser, David H. Moreau, and Bruce Stiftel

Irrigation Management in Developing Countries, Kenneth Nobe and R. K. Sampath

Also of Interest

Public Representation in Environmental Policymaking: The Case of Water Quality Management, Sheldon Kamieniecki

Water Needs for the Future: Political, Economic, Legal, and Technological Issues in a National and International Framework, edited by Ved P. Nanda

Natural Resource Economics: Selected Papers, S. V. Ciriacy-Wantrup, edited by Richard C. Bishop and Stephen O. Andersen

About the Book and Editor

Efficient energy and water management is critical in semiarid areas that need irrigation for crop and livestock production. This book focuses on the prospects of using irrigation in agricultural production as energy becomes increasingly scarce. Economic and technical issues that affect irrigation-management decisions are considered, with emphasis on the amount of energy required in the use of irrigation and on adjustments made in agricultural production when the cost of energy is high. Alternative energy forms for irrigation pumping--especially wind and solar power--are assessed. Since livestock production in most of the western United States depends heavily on water, the contributors compare the energy intensity of livestock production in irrigated areas with that of range livestock production in nonirrigated areas. In addition, the contributors examine potential adjustments in agricultural production in response to energy and water shortages and evaluate the relative importance of irrigation as one use of water and energy. The findings have applications for irrigated areas throughout the world.

Norman K. Whittlesey is professor in the Department of Agricultural Economics, Washington State University, Pullman, Washington.

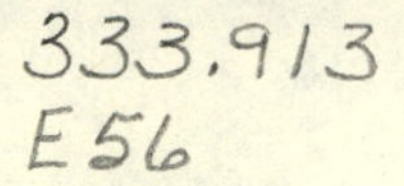

Energy and Water Management in Western Irrigated Agriculture

edited by Norman K. Whittlesey

Studies in Water Policy and Management, No. 7

Westview Press / Boulder and London

Studies in Water Policy and Management

Published in 1986 in the United States of America by Westview Press, Inc.; Frederick A. Praeger, Publisher; 5500 Central Avenue, Boulder, Colorado 80301

Library of Congress Catalog Card Number: 85-52184
ISBN: 0-8133-7086-8

Composition for this book was provided by the editor.
This book was produced without formal editing by the publisher.

Printed and bound in the United States of America

The paper used in this publication meets the minimum requirements of the American National Standard for Permanence of Paper for Printed Library Materials Z39.48-1984.

6 5 4 3 2 1

To Cynthia

Contents

Tables and Figures

Tables

Figures

Foreword

This volume on energy and water conservation practices is the seventh in Westview's Water Policy and Management series. The study assesses the future of irrigated agriculture in arid areas in the face of increasing energy and water scarcity. Conservation practices are described in terms of technology and economic feasibility. Alternative sources of energy for pumping are evaluated, as is a range of conservation techniques and adjustments: optimal irrgation system design, management practices, cropping patterns, and livestock production practices. The book will be valuable to all agricultural economists and will serve as a handbook to extension personnel and consultants.

Charles W. Howe
General Editor

Charles W. Howe is professor of economics at the University of Colorado, Boulder, Colorado. He was director of the Water Resources Program at Resources in the Future, Inc. and has been a teacher and a consultant in several Third World countries. His books include Benefit-Cost Analysis for Water System Planning, Natural Resource Economics, *and* Management of Renewable Resources in Developing Countries.

Preface

The energy shortages and embargoes of the early 1970s had profound effects on agriculture. Energy had grown to become a major input in agricultural production, processing, and marketing. However, because it had always been cheap and plentiful, little was known about how much energy was being used in agriculture for food production, processing, or marketing. There were no answers to policy makers' questions about how agriculture might adjust to energy shortages or dramatic price increases and how food production and costs would be affected by the changed energy situation.

An insidious problem that was noticed after the first round of effects had been felt, was the impact of higher energy costs on irrigated agriculture. Irrigating requires energy to lift water from rivers or from the ground, to pressurize sprinkler systems, and even to pump water in those farming areas that must be drained after they have been irrigated. This has always been true but only recently recognized as a major problem. With the rapid increase of energy prices, there emerged a new set of questions about irrigated agriculture. What part of the 35 million irrigated acres in the West requiring energy for pumping would continue in irrigated farming? Would there be a drastic change in the composition of crops grown on these acres? Would farm income be reduced drastically? Should public policy be used to provide special incentives for irrigated agriculture? What changes in technology might help farm operators or irrigation districts reduce their consumption of water? These and scores of other questions were addressed as the price of energy continued to rise and as the changing relative cost of energy forced western irrigated agriculture to consider major alternatives.

Researchers at several agricultural experiment stations in the West and with the United States Department of Agriculture began to inquire into these and other questions related to the use of energy in western agriculture. Eleven states and the USDA (Alaska, Arizona, California, Colorado, Hawaii, Idaho, Nevada, New Mexico, Oregon, Texas, and Washington) organized to attack these problems under the auspices of Western Regional Research Project W-140, "Energy in Western Agriculture - Adjustments, Alternatives, and Policies." This book is the production of 10 years of coordinated research under this project that related to the energy problems of irrigated agriculture. This is Western Regional Publication Number 006.

The book is written for the layman, teacher, or policy maker who wants to know more about the technology and management of irrigated agriculture. Its particular emphasis is the variety and complexity of the interactions between energy and water. It will be a useful reference or supplemental text for students of energy and water management in arid regions. While the location of research reported in the book is the western United States, most of the discussions have applications to modern irrigated agriculture anywhere in the world. It is an important contribution to the subjects of energy and water use in irrigated agriculture.

Norman K. Whittlesey

Acknowledgments

Special thanks are due to Sue Osterback who patiently and expertly typed the material for this book and to Sharon Baum who helped to finish the final typing. It could not have been accomplished without their help.

The committee members of Western Regional Research Project W-140 were very supportive in this effort. Authors were timely in developing their contributions and generous in their acceptance of suggestions for change.

Finally, the administrative support that allowed and encouraged the book to be written is recognized. Loy Sammet and Dennis Oldenstadt are singled out for the helpful administrative guidance provided the W-140 Committee during the past 10 years.

N.K.W.

1
Problems, Concepts, and Policy Issues

Norman K. Whittlesey

INTRODUCTION

This chapter primarily serves to introduce the remainder of this publication. Its purpose is to raise questions about the interactions of energy and water use in irrigated agriculture, but not to answer these questions with any extended discourse or data analysis. The following chapters will each address some part of irrigated agriculture in a way that will sharpen the focus on the role of energy in this important segment of agriculture.

This is not to be a comprehensive book about irrigation, irrigation management, water resource policy, or energy policy. However, in a sense, it is about all of these things. It is intended to highlight some of the problems, concepts, and policy issues related to energy use in irrigated agriculture. We hope to sort myth from reality regarding the use of energy and water to produce food and fiber for U.S. and world consumers.

The area of geographic focus for this book is the western half of the United States. However, many of the technical and managerial concepts or problems described could be extended to any part of the world with similar circumstances. In the end, we hope to give the reader or student of this subject a better understanding of the real world, to help remove interest from those areas without significant problems, and to increase attention on those areas requiring future research or policy action. To achieve this goal it will be necessary to delve into numerous areas of irrigation management and technology. We will, at various times, describe irrigated agriculture in general terms or discuss energy policy and water resource policy as a means of dealing with the future resource problems of irrigated agriculture.

AGRICULTURAL PLANNING NEEDS

Past policies of water resource management and cheap energy have attracted irrigation into many areas requiring high pump lifts from both ground and surface water sources while also encouraging the adoption of energy intensive technologies in all areas of irrigated agriculture. The result has been to leave some segments of irrigated agriculture vulnerable to rising costs of energy and potential shortage. To survive and compete in the future, irrigated agriculture must adopt many changes in water use, energy use, irrigation management, cropping patterns, and irrigation technology.

Energy Implications

The oil embargoes of the 1970s created a concern about the planning needs of irrigated agriculture for pumping energy. Similarly, a shortage of water in the Columbia River system can create a shortage of hydropower for all users of electricity in the region. While the possibility of brief energy shortages, one month to a year, for irrigation pumping seems rather remote as this is being written, events of the recent past will tell us it is a potential threat. There are numerous implications of an energy deficiency. If adequate water cannot be delivered to crops in a timely fashion, the losses can be severe. For annual crops already planted the loss may range from decreased yield to a complete crop failure. For perennial crops such as orchards and vineyards, the lack of water could impose plant damage with much longer lasting effects and greater economic cost. Of course, the impact of energy shortages will be influenced by such things as duration, length of prior warning, and the plans for dealing with a shortage.

In those areas where there is a potential for short-run energy supply interruption there is a need for contingency planning to deal with the problems imposed. This discourse is not intended to alarm but does suggest a need for research and policy planning to deal with such problems as they arise or, more important, to avoid them entirely.

In the long-run, the energy problems of irrigated agriculture are likely to be more certain, as well as complex, pervasive, and potentially damaging. However, the opportunities for dealing with these problems are abundant, but waiting to be exploited by managerial, research, and policy actions. There will be shifts in regional comparative advantage in agricultural production due to increasing energy costs. These shifts will lead to changing cropping patterns and, in some cases, significant changes in agricultural production. For some irrigated areas of the West, the major decisions will be about how to deal with the economic and social impacts of changing levels of agricultural activity or income. It is important to remember, however, that production patterns are determined by regional comparative advantage, not absolute advantage in cost of production. Moreover, the value of fixed assets within a region will largely reflect the impacts of increasing the cost of energy as a variable production input. That is, land values will absorb the changing return to fixed resources.

Those areas of the West that have traditionally been exposed to high costs of energy due to high pump lift conditions or high prices for energy have already absorbed many available technological and managerial options for minimizing energy costs. Those areas will have fewer available options for future adjustment than others which have yet to experience significant increases in real energy costs. Nevertheless, the long-run adjustment opportunities for all irrigated areas dependent upon energy for pumping water are still numerous. Management can improve pump performance and water management. Crop selection will reflect subtle shifts in net income potential as influenced by water requirement. Many technological changes are underway and more are on the horizon. In short, the irrigated areas of the West are more energy intensive in crop production than the non-irrigated areas of the nation. Future changes in energy policy, supply, or cost will differentially affect these areas. However, most will adjust and survive.

Water Implications

Throughout the arid West, water is generally the limiting factor of production. Natural rainfall will not support the levels of crop production held in the potential of soils and other climatic factors. In fact, areas with the warmest climate and longest growing seasons are frequently the most deficient in water supply. Hence, the need for irrigation arises.

Water is generally a heavy, and difficult resource to capture and transport to the land. Where possible, gravity flow systems have been developed to move the water to land. Many areas originally serviced by gravity flow application systems have adopted sprinkler technologies to increase crop productivity or stretch scarce water supplies, but increasing the energy dependence of irrigated agriculture. All lands irrigated from groundwater and many from surface water require pumping for delivery as well as application.

Many irrigated areas of the West rely upon diversions from surface streams or storage dams. In these cases water is a flow resource with a limited seasonal supply. In most such areas there is no surplus or unused water. Irrigators are normally provided a fixed amount of water per acre or per farm within a growing season. Frequently the amount of available water determines the type, mix, and productivity of crops that are grown. To add more water to presently irrigated land or develop additional irrigated land would be impossible without decreasing other diversion or instream uses of water.

Irrigated agriculture is the largest consumer of water in the West, consuming nearly 90 percent of all water actually transformed or evapotranspirated from the liquid form. However, it is not necessarily the largest or only user of water. Municipal, industrial, power generation, recreation, and fisheries are some of the major competitive users of water in the West. To increase agricultural uses of water, particularly surface water sources, in any area will require decrease of one or more competitive uses.

The competitive users of water frequently place higher economic value on the water than does agriculture, making it possible to bid water away from agriculture when that is necessary or desirable. However, the legal rights for water use afforded agriculture throughout the West generally provide irrigation much greater protection than the in-situ uses of power generation, recreation, wildlife, or fisheries. There are strong needs in many areas for institutional and legal changes to allow for better allocation of water among competing uses to meet societal goals. Agriculture must be prepared to meet the increasing competition for scarce water resources by becoming more efficient in its use.

In many irrigated areas groundwater is the major source of supply. Most of these aquifers have some natural recharge, but frequently the rate of recharge is less than the rate of withdrawal for irrigation and other uses. Groundwater mining is occurring in large areas of the Great Plains, Arizona, California, and Washington.

These areas where irrigation depends heavily on groundwater mining face varying degrees of water scarcity in the future. To continue present rates of use will eventually deplete the resource or put the costs of water acquisition beyond the reach of agriculture. In some areas agriculture is already on the decline while in other areas the real problems of adjustment are farther away in the future. As water tables fall and pump lifts increase the farmer faces larger energy costs for lifting the water from beneath the ground. Also, the attempts to use scarce water supplies more efficiently frequently requires a greater dependence on energy intensive irrigation technologies.

Water is generally limited in quantity where used for irrigation. It is not available to substitute further for other resources such as land, labor, or energy. To face the rising energy cost and water scarcity problems of the future will require irrigated agriculture to continue seeking better managerial and technological options for its use.

ENERGY DEPENDENCE IN IRRIGATION

Energy and water use in irrigation are closely intertwined. Farmers who once accepted energy use as necessary, but managerially unimportant due to its abundance and low cost, can no longer do so. However, changes in recent years dictate that energy must be managed with the same amount of skill and intensity as the water resource.

Energy in its broadest terms may be an unlimited resource. However, as a practical matter, energy for lifting and applying water to land for crop production faces definite supply limits. Moreover, agriculture is in competition with other segments of the economy for use of this energy and is frequently at the mercy of policies or prices set by or for these competitors. The economic viability of large segments of irrigated agriculture is dependent upon the availability and cost of pumping energy.

In many cases survival will depend upon abilities to conserve energy use, substitute other factors for energy, or increase the productivity of energy used.

Heterogeneity of Irrigated Agriculture

There are large portions of the irrigated West where water diversion, delivery, and application are accomplished entirely through gravity flow systems. These areas range from streamside alpine meadows that are flood irrigated with spring runoff waters to complex water storage and delivery systems that move the water hundreds of miles before applying it to high value crops through very efficient irrigation systems. These areas are dependent upon the use of energy for water delivery or application and, hence, are not particularly vulnerable to problems of rising energy costs or potential shortage. In fact, some of them may benefit from increased comparative advantage in production location as competing irrigated regions suffer from problems associated with energy use. In any case, it must be recognized that "irrigated agriculture" is not a monolithic entity to be uniformly affected by future changes in energy supply, cost or policy.

Within the portion of irrigated agriculture that does use energy for pumping irrigation water, there is a wide diversity in the level of energy employed. The reasons for this are equally diverse. Many farming areas have voluntarily converted from gravity flow application systems to various forms of sprinkler application. In some cases this has been a substitution of capital and energy for labor to reduce labor acquisition and management problems. Sometimes it has been done to improve the efficiency of water use and raise crop productivity. These situations generally use relatively small amounts of pumping energy per unit of production and continue to have the option of returning to gravity flow methods of irrigation if energy use becomes prohibitive. Of course, changes in method of irrigation are not without cost. Each situation must be investigated for its adjustment opportunities and vulnerability to energy supply problems.

At the extreme of energy intensity are those farms required to lift water over large elevations. To make the most efficient use of such water, it must also be applied through pressurized sprinkler systems. Farms in this situation are most vulnerable to changing energy supply conditions. While many adjustments to changing factor price ratios are possible, it is not possible to avoid energy use without the cessation of irrigated farming.

The source and cost of energy for irrigation pumping also ranges widely throughout the West. In Texas and Arizona pumping has relied heavily upon natural gas while the Pacific Northwest almost exclusively uses electricity. Other regions depend upon diesel or combinations of all these sources.

Different parts of irrigated agriculture are variously affected by the type of energy used for irrigation. We have been through periods of temporary shortage for petroleum products. Natural gas prices have been deregulated and allowed to increase significantly. The Pacific Northwest no longer enjoys the comfort of a perpetually low electricity price and is feeling the effects of investments in some very expensive nuclear power plants. In all cases, the era of cheap and abundant energy for irrigation seems to be ending.

Energy Policy and Market Imperfection

Some of the regional differences in energy use and dependence have developed in response to past policies and market imperfections that tended to keep energy prices of all forms below their real social value or cost. Cheap energy in the past, partially due to government subsidies for energy development or price regulation, encouraged energy intensive irrigation development in many parts of the West. There are many thousands of acres that are irrigated from both ground and surface waters with pump lifts exceeding 500 feet. The overdraft of the Ogallala aquifer in Texas was encouraged by cheap, price regulated natural gas. Cheap electricity from federal dams on the Columbia River enticed irrigation into areas of very high pump lift in the Pacific Northwest.

The significant shifts in energy policy and energy cost of recent years is now causing serious problems for some irrigated areas. The problems are particularly acute for high pump lift situations where energy use cannot be avoided but only adjusted marginally by changing crop water use or irrigation method. These are the farms that will be impacted most by the rising real costs of energy. Shifts in energy cost do not move uniformly through time, nor do the prices of different energy forms move together by responding to the same phenomena. Different segments of irrigated agriculture are impacted differently over time.

Most management decisions are profit motivated and respond only to current or expected market prices for inputs and products. To the extent that legal or institutional constraints or market imperfections cause the choice indicators (prices) to be distorted from true social values, it is possible for misallocation of resources to occur. The uncertainties regarding future energy supply and demand probably do result in inaccurate prices for energy. The variations in national policy regarding energy pricing, technology development, and supply regulation reflect these uncertainties. Irrigated agriculture, like other sectors of the economy, must continually adjust to the changes in energy prices and availability.

There is not a single energy problem in irrigated agriculture to be solved with a single monolithic action. Instead there is a vast multitude of energy-water relationships throughout the West. Each circumstance presents its own degree of problem difficulty and cries for its own particular solution.

TECHNOLOGICAL CONSIDERATIONS

Technology Development and Adoption

Irrigation technology is continually being improved. Manufacturers and researchers are providing a steady stream of new technology for irrigated agriculture and the farmers themselves frequently contribute to the process. Irrigation technology can be divided between factor saving and output increasing types, among many other classifications. Both of these can affect the output/energy-water ratios that are so important to agriculture. Most of the technological developments of the past have responded to policies of cheap energy by being capital and energy using and labor saving. In some cases, the improvements in water use efficiency have managed to hold down or reduce the relative costs of energy, despite rising energy prices. Technology development and adoption will play major roles in the survivability of irrigation in many areas as the energy picture unfolds into the future.

New technologies must be adapted to site specific characteristics of soil type, slope, water quality, water quantity, climate, crop selection, pump lift, and energy cost. Adoption of new technology is influenced by capital cost, capital availability, economic feasibility, risk aversion, labor and management inputs, and the fixed investments in present technologies. Energy costs must rise significantly in real terms to cause farmers to abandon present investments for new energy saving technologies. More frequently in the past the energy intensity of irrigation technology has increased as adoption has occurred. Energy and capital have substituted for labor and water while increasing crop productivity.

The adoption of irrigation technology continues to be based on criteria of economic efficiency, not energy or water efficiency. It is only fortuitous if these phenomena should be simultaneously achieved. Water scarcity has always been a concern in western irrigated areas. Energy is now looming as an important and costly input to irrigated agriculture. Technology development and adoption in the future will be increasingly influenced by the cost and future prospects for energy. However, it is expected that decisions regarding selection of new technology will continue to be based on economic efficiency without special regard for energy. Irrigation systems that minimize energy use generally do not yet meet the test of economic efficiency and would not be expected to do so unless real energy costs increase many fold.

Opportunities for Factor Substitution

A major policy and managerial question relates to the potential substitution among water, energy, and other factors of production in irrigated agriculture. Economic theory dictates that where such opportunities do exist there will always be adjustments in factor shares (relative proportions) in response to changing factor price ratios.

Recent history has generally shown energy being substituted for labor, water, and other scarce production inputs in irrigated agriculture. This process has sometimes continued during periods when the real costs of energy were increasing faster than other input costs. However, it must be noted that irrigation technology development and adoption occurs for many reasons, including risk aversion, labor management, environmental concerns and increased productivity, and cost minimization.

A new irrigation technology may be adopted for its potential to increase the productivity of land, water, or fertilizer used in crop production. This may be accomplished by reducing water runoff, deep percolation, or evaporation losses. Frequently such technologies may increase the intensity of energy use but the increased productivity of other resources will still increase net farm income. A technological shift toward the use of more energy without offsetting reductions in other factor needs or increasing crop output may not increase net farm income. However, it may reduce income variability or environmental damage due to soil or nutrient losses.

Simple changes in factor price ratios may not always dictate the future of factor shares in this industry. In fact, the energy intensity of irrigated agriculture is likely to increase in many areas despite the increasing real cost of energy. However, there are opportunities for shifting to technologies that can reduce current energy use levels when that need is apparent. Irrigated agriculture is not about to disappear for lack of energy or due to its increasing cost.

CONCLUSIONS

This brief view of irrigated agriculture and its impending future serves to introduce the remainder of this book. Irrigated agriculture is declining in some areas of the West due to urban development, declining water supplies, and rising energy costs. However, irrigation development continues in other areas where water supplies and economic conditions will permit its growth. Projections of 40 or 50 years into the future indicate that irrigation will continue to grow and be more important in western agriculture than it is today. But there will continue to be many areas of painful adjustment as water and energy supply constraints press upon portions of irrigated agriculture. Irrigated agriculture will have to adjust but it will survive. The remainder of this book is about how many of the battles will be fought.

REFERENCES

English, M.E., R.H. Cuenca, K.L. Chen, R.B. Wensink, and J.W. Wolfe. Analysis of Energy Used by Irrigation Systems. Agricultural Engineering Department, Oregon State University, Corvallis, 99 pp., 1982.

English, Marshall, Roger Krognich, and David Eakin. "Potential Conservation of Energy and Water." In An Analysis of Agricultural Potential in the Pacific Northwest with Respect to Water and Energy, Northwest Agricultural Development Project, Dec. 1979.

Jensen, Marvin E. "Overview--Irrigation in U.S. Arid and Semi-Arid Lands." In Water-Related Technologies for Sustaining Agriculture in U.S. Arid and Semi-arid Lands. Office of Technology Assessment, Oct. 1982.

Whittlesey, N.K., J. Buteau, and W.R. Butcher. Energy Tradeoffs of Irrigation Development in the Pacific Northwest. Washington Agricultural Experiment Station Bulletin 0896, 1981.

Whittlesey, N.K. "Energy and Irrigated Agriculture: Problems and Prospects." Proceedings, Agricultural Conference Days, Corvallis, Oregon, March 1983.

Whittlesey, N.K., et al. "Demand for Electricity by Pacific Northwest Irrigated Agriculture." Report to Bonneville Power Administration, January 1982.

2 The Historical Setting for Irrigation in the West

Daniel J. Bernardo
Norman K. Whittlesey

INTRODUCTION

Agricultural development has played a crucial role in the economic growth of the western United States in the twentieth century. An important factor contributing to this development has been the ability of the region to use its scarce water supplies for irrigating crops. By 1977 over 60.7 million acres throughout the United States were being irrigated (USDA/SCS) and 50.2 million were located in the 17 western states. Since rainfall in many western regions is inadequate or too irregular to support agricultural production, western agriculture depends heavily upon irrigation. Approximately 20 percent of the agricultural land in the West is irrigated, while irrigation accounts for only 3.5 percent of agricultural land in the 31 eastern states.

In the Southern Plains and Northern Plains irrigated land accounts for about 9 percent of the 173.8 million acres in agricultural production. In the Pacific and Mountain Regions, however, irrigation accounts for 44 and 64 percent, respectively, of the agricultural land.

Despite the success of irrigation in the western United States, its future is uncertain. Water shortages and increasing energy costs are affecting many irrigated areas and, though irrigation is not disappearing, the problems of adjusting to these changing conditions are sometimes difficult.

Even though water scarcity has been in evidence for several decades, response of irrigation has been limited. The historic method of augmenting supplies through groundwater mining, surface storage, or interbasin transfers was the common response to impending water scarcity. These solutions may be infeasible for increasing irrigation or solving future water shortage problems since most surface and groundwater sites have been fully exploited. In addition, the environmental consequences of water project development are being given greater weight in policy decisions. These changes have begun to impact the rate, and in some regions the direction, of irrigation growth in the western United States.

WESTERN IRRIGATION DEVELOPMENT

Early Development and the Reclamation Era

The growth of irrigated acreage in the 17 western states has not been uniform through time or among regions. Political, economic, and institutional forces have all interacted to encourage and sometimes inhibit irrigation development. Basically, western irrigation development can be divided into two distinct eras. The period from 1870-1900 was primarily characterized by private development. The twentieth century, on the other hand, has been heavily influenced by various pieces of federal land and water policy legislation.

Irrigation began in the western United States to provide early settlers with food and fiber. Prior to 1850, with the exception of the Mormon development in Utah, no large amounts of irrigated acreage existed in the western United States. Most irrigated acreage was found in small tracts of land along rivers and streams. After these early developments, limited commercial growth occurred in a number of small, fertile areas in several western states, especially California. In 1870, the estimated irrigated land in the West did not exceed 300,000 acres. Private development was speculative and frequently financed through the sale of stocks or bonds. Consequently, growth was restricted to the low-cost development along river bottoms. By 1890, there were more than 3.6 million acres of irrigated land in the West, primarily financed by individuals and farmer associations (Golze).

Eventually, development began to depend upon support from the federal and state governments. Development costs exceeded the financial and organizational capability of private speculators and associations. Three major pieces of legislation enacted in the late 1800s spurred twentieth century irrigation growth--the Homestead Act, the Desert Land Act, and the Carey Act. Coupled with the Reclamation Act of 1902, these three doctrines provided the legal framework for much of the government involvement in western irrigation development.

The Homestead Act (1862) provided the foundation for settlement and agricultural development in 16 of the 17 western states (Golze). Although the act did not have an immediate direct effect on the amount of western irrigated acreage, it did significantly impact western agricultural growth and subsequent irrigation development. The Homestead Act permitted any person over 20 years of age to select 160 acres in the public domain and to acquire its title after residing on it for 5 years and completing cultivation requirements. The residency requirement was then amended to 14 months to further stimulate interest in western settlement.

The Desert Land Act (March 3, 1877) provided that title to 640 acres of arid land could be obtained by conducting water upon it, reclaiming the land within 3 years, and by paying the government $1.25 per acre (Golze). This legislation was passed in recognition of the limited application of the Homestead Act to the more arid lands of the West. The act was the first major effort by the federal government to encourage irrigation and cultivation of the vast expanses of arid land in the West. Over 10,000,000 acres were patented under the Desert Land Act, all for the purpose of private development. Unfortunately, various weaknesses in the text of the act led to speculation and title attainment without fulfilling the proper requirements. The resulting public pressure led to the drafting of the Carey Act.

The primary objective of the Carey Act was to aid public-land states in the reclamation of desert lands. The Carey Act granted to each state an amount not exceeding 1,000,000 acres and directed states to induce investment to stimulate irrigation (Golze). Most success of the Carey Act was experienced in Idaho and Wyoming; both states received additional acreage allotments for distribution. On the whole, however, the Carey Act was not successful in the majority of participating states.

Although nineteenth century land policy legislation did not provide the agricultural growth anticipated during their conception, several million acres were developed as a consequence of such enactments. Both the limited financial resources of state governments and apparent lack of interest in some states were major reasons for the limited success of these programs. Though limited, the impact of this legislation was continuing to be felt throughout the early 1900s. Another important outgrowth of this legislation was the development of a legal and institutional framework at both the federal and state levels capable of instituting reclamation law. Even today some land in southern Idaho is being developed under influence of the Desert Land Act and the Carey Act.

The Reclamation Act of 1902 is the foundation of modern reclamation in the United States and serves as the impetus for federal development of irrigation and hydroelectric power in the United States. Since 1902, most federal enactments for reclamation have been interpretations of the act or modification of its requirements. It authorized the Secretary of the Interior to investigate the construction of irrigation works for storage, diversion and development of water. Another important provision established the Reclamation Fund for the purpose of financing irrigation works. In addition, the Reclamation Service was created as the capital authority for the administration of reclamation activities (Ostrom). Essentially, the Reclamation Act of 1902 served to formalize the legal framework in which federal participation in water resource development was to occur.

As a result of the Reclamation Act of 1902, a tremendous growth in lands receiving either full or supplemental irrigation water through federal projects was observed. Land receiving "federal water" grew from 400,000 to 3,000,000 acres from 1910 to 1930 (Golze). By 1980, over 11,000,000 acres were estimated to be receiving either full or supplemental irrigation water from federal works. In more recent times, the role of the U.S. Bureau of Reclamation (USBR) has expanded to provide water for municipal and industrial use, hydroelectric power, water channels for navigation, flood control, and recreation facilities. It has been estimated that over $6 billion has been spent on water related projects by the USBR since its inception (Ostrom).

Much of the growth experience since 1900 has resulted from some important amendments to the initial reclamation legislation. It was recognized early that certain budgetary stipulations in the 1902 legislation severly restricted its application. As a result, two pieces of legislation were passed to loosen the payback requirements of the original law. The 1914 Reclamation Extension Act increased the repayment period from 10 to 20 years. Later amendments have extended repayment periods even further. In addition, the Reclamation Project Act (1939) specified that irrigators were only responsible for the debt they are able to repay (Ostrom). Consequently, despite original stipulations that user charges be sufficient to recover all construction costs, cost recovery has not generally occurred. Much of the aforementioned $6 billion dollars has been a public subsidy to agricultural development.

Regional Growth Rates

The cumulative effect of the federal land policy and reclamation legislation discussed above was a continual expansion of irrigated acreage in the western United States throughout the first half of this century. With the assistance of the federal government's reclamation effort, total irrigated acreage in the West expanded to more than 20 million acres by 1940. Of this amount, an estimated 3,800,000 acres were irrigated with water furnished by the government; the remaining 16,600,000 acres were developed by private enterprise (Golze).

Following World War II, the trends in western irrigation development began to change. A long-run view of irrigation expansion in this era using agricultural census data is presented in Table 2.1. These data indicate that although irrigated acreage has continued to grow, the rate of growth has been sporadic and generally declining. For example, during the period from 1945 to 1950 the annual rate of growth was 4.5 percent. In contrast, the growth rate in the 1969-1974 period was only 1.1 percent. Although a number of reasons contribute to this phenomenon, one simple explanation is the decline in available water supplies. The increase in growth rate shown in the 1974-1978 period is due primarily to acreage expansion in California and the Ogallala region of the Great Plains. Subsequent to 1978 there has been very little irrigation growth anywhere in the West.

Irrigation growth rates have also differed greatly among regions during the 1945-1978 period. Table 2.2 demonstrates the change in irrigated acreage for each of the West's four production regions in the 1945-1978 period. During this period, the Northern and Southern Plains accounted for 31 and 34 percent respectively of the total

Table 2.1
Growth of irrigated acreage in the West, 1945-1978

Year	Change in Acres Irrigated (1,000,000)	Average Annual Growth of Acres Irrigated (%)
1945-1950	4.8	4.5
1950-1954	2.7	2.7
1954-1959	3.9	2.7
1957-1964	2.5	1.6
1964-1969	1.6	0.9
1969-1974	1.8	1.1
1974-1978	7.0	4.5

Source: U.S. Bureau of Census, Census of Agriculture, various years.

expansion of irrigated acreage. In contrast, the Mountain and Pacific Regions accounted for only 12 and 23 percent of the total growth.

Historic irrigation acreage in the U.S. varies significantly depending upon the data source used. Agricultural Census and Natural Resource Inventory are the two primary data sources utilized. Agricultural Census data is thought to continually underestimate total acreage; however, this should not significantly impact rates of change.

This regional growth has by no means been spread evenly across time. Prior to 1945, growth was concentrated in the Mountain and Pacific Regions. In 1945 the Mountain Region accounted for over half of the western irrigated acreage. In the first ten years of the post-war era (1945-1954), the Southern Plains and Pacific Regions dominated irrigation growth. The Pacific Region was spurred by the development of large water projects such as the Imperial Valley and Central Valley projects in California. The Southern Plains' growth was primarily centered in the High Plains Region of Texas and Oklahoma which overlies the Ogallala groundwater formation.

In the 1955-1964 period, growth was relatively uniform throughout the four regions. Great variation among the

Table 2.2
Irrigated acreage by state and production region

Region-State	1944	1954	1964	1974	1978
	(thousands of acres)				
Pacific:					
California	4952	7048	7599	7749	8603
Oregon	1129	1490	1608	1561	1920
Washington	520	778	1150	1309	1691
SUBTOTAL	6601	9316	10357	10619	12214
Mountain:					
Arizona	736	1177	1125	1153	1211
Colorado	3699	2263	2690	2874	3458
Idaho	2062	2325	2802	2859	3508
Montana	1555	1891	1893	1759	2086
Nevada	674	567	825	778	899
Utah	1124	1073	1092	970	1185
New Mexico	535	650	813	867	904
Wyoming	1334	1283	1571	1460	1685
SUBTOTAL	11719	11229	12811	12720	14936
South Plains:					
Oklahoma	2	108	302	515	602
Texas	1320	4707	6385	6594	7018
SUBTOTAL	1322	4815	6687	7109	7620
North Plains:					
Kansas	96	332	1004	2010	2686
Nebraska	632	1171	2169	3967	5698
S. Dakota	52	90	130	152	341
N. Dakota	23	38	51	78	141
SUBTOTAL	803	1631	3354	6207	8866
TOTAL	20445	26991	33209	36655	43636

Source: U.S. Bureau of Census, Census of Agriculture, various years.

four regions was again the norm in the 1965-1978 period. Due to expanded groundwater use, the Northern Plains Region accounted for over 82 percent of the period's acreage expansion. Over 25 percent of the western irrigation expansion during the 1974-1978 period occurred in Nebraska (Bureau of Census).

It is important to note that irrigation cannot be credited for the entire output produced on irrigated acreage. Much of these lands could produce a crop without the application of irrigation water. This is especially true for the land developed recently in semi-arid regions of the Great Plains. Much of this land was in some form of grain production prior to irrigation development.

Impact on Yield and Cropping Patterns

Simply considering the acreage of irrigated land does not accurately reflect the importance of water resource development to western agriculture. Irrigation has expanded the number of acres under cultivation, but it has also impacted the yield and cropping patterns of irrigated lands.

Table 2.3 seeks to illustrate the effect of irrigation on crop yield by comparing average yield for selective crops on nonirrigated land, partially irrigated land; and

Table 2.3
Yield per acre of selected crops grown in the West

Crop	Non-Irrigated	Partly Irrigated	Fully Irrigated
Corn for Grain (bu)	94.9	97.0	114.3
Sorghum for Grain (bu)	48.1	49.6	71.3
Wheat (bu)	28.5	29.7	54.7
Barley (bu)	42.7	45.1	67.5
Dry Beans (cwt)	10.6	13.6	16.8
Cotton (bales)	0.7	0.6	1.2

Source: Jensen.

fully irrigated land. Obviously, not all of the yield difference can be attributed to irrigation. These figures are averages for western agriculture and the effects of climate, region, and irrigation may be confounded since irrigated crops are not always grown in the same region as nonirrigated crops. Nonetheless, it is apparent that irrigation does have a dramatic impact on crop productivity. Average crop yields for major crops on irrigated lands in the United States range from 16 to 92 percent more than on non-irrigated land (Jensen).

Little change has occurred in the proportion of irrigated acreage allocated to most crops in the past 30 years. Cotton, hay, vegetables, and orchards have all remained fairly constant in their relative shares of irrigated land. In 1978, the West's irrigated acreage was allocated 45 percent to grains, 24 percent to hay, 7 percent to cotton, 13 percent to vegetables and orchards, and 11 percent to all other crops. The increasing importance of grain can be primarily attributed to corn which has risen eight-fold in irrigated acreage from 1950 to 1978. Most of this irrigated corn is in the Northern Plains Region on lands that previously produced corn without irrigation. The irrigation in this case increases average crop yield and reduces the risk of weather variations.

Value of Irrigated Products

In 1978 an estimated $11.4 billion worth of crops were grown under irrigation in the 17 western states. In addition, a large portion of the $26.8 billion of livestock products raised in the West was derived from irrigated farms. Table 2.4 shows the acreage and value of products raised on irrigated western land. These crops represent over 23 percent of the total value of agricultural crops grown in the United States while using only 9 percent of the nation's cropland.

Using only 6.3 percent of irrigated land, orchards contributed an estimated $1,789 million in value of production in 1978 or 15.6 percent of the total. In comparison, hay crops used 24.5 percent of total irrigated lands to contribute 9.0 percent of crop value. However, this hay production has a complementary relationship with much of the range livestock production in the West. This allows the grazing lands to be much more productive than would be

Table 2.4
Acreage and total value of crops grown on irrigated land (1978)

Crop	Acreage	Percent of Total Acreage	Average Yield	Total Value
	(1,000)			million $
Corn Grain (bu)	7,849	21.4	102.1	1,619.8
Corn Silage (T)	1,255	3.4	16.8	425.2
Sorghum Grain (bu)	2,019	5.5	70.6	276.5
Sorghum Silage (T)	111	.3	16.8	37.2
Wheat (bu)	2,987	8.1	45.6	400.4
Oats (bu)	222	.6	56.0	14.7
Barley (bu)	1,937	5.3	61.6	226.7
Rye (bu)	4	[a]	29.3	.2
Rice & Other Small Grains (cwt)	1,191	3.3	44.9	412.8
Soybeans (bu)	442	1.2	28.45	83.5
Peanuts (lbs)	169	.5	2443.0	86.9
Cotton (lbs)	2555	7.0	700.0	1,055.0
Potatoes (cwt)	715	2.0	274.0	697.8
Sugar Beets (T)	764	2.1	21.4	395.8
Dry Field Beans (cwt)	468	1.3	10.3	78.6
Sweet Potatoes (cwt)	8	[a]	116.0	9.5
Hay Crops (T)	8,954	24.5	2.3	1,029.7
Alfalfa Seed (lbs)	202	.6	440.0	107.1
Clover Seed (lbs)	21	.1	110.0	1.6
Other (lbs)	74	.2	14.6	13.2
All Vegetables	1,447	4.0		1,672.4
Land in Orchards	2,305	6.3		1,789.0
Berries (lbs)	29	.1	178.0	158.4
Nursery Products	115	.3		541.1
Other Crops	581	1.6		232.2
TOTAL				11,436.3

Source: U.S. Bureau of Census, Census of Agriculture.

[a]Less than 0.1 percent.

possible without the support of the irrigated forage production.

LEGAL AND INSTITUTIONAL CONSTRAINTS ON IRRIGATED AGRICULTURE

The relative abundance of water supplies and ease of irrigation development which characterized the earlier periods of western agricultural development are no longer the norm in most regions. A variety of forces have contributed to disruption of traditional water resource policy. Surface water is scarce and has competing uses while groundwater supplies are generally declining. Environmental concerns have become more prominent in water resource management. Water projects have been a popular target in the political budget slashing activities of the recent past. Finally, the public subsidies provided to irrigators by water projects have been heavily criticized by politicians and economists. These forces, among others, have acted to disrupt irrigation expansion in the western United States.

The institutional and legal framework in which the current water allocation system operates significantly affect the management and policy decisions that can be implemented. Each irrigation setting is influenced by a unique set of interrelated private and public institutions, as well as a complex legal framework.

Most of the laws and institutions governing water use and allocation came into existence during a period when expansion and exploitation were major goals of water resource policy. In many cases, the institutional structure has been found to limit rather than facilitate conservation and efficient use of water. Some proposed solutions to existing water scarcity problems, such as market-oriented allocation systems, will require significant modifications of current institutions. Administrative and political institutions will have to be created or modified to encourage conservation and transfer of the resource. Most impacted will be the current system of water rights governing the allocation of water for agricultural, municipal, and industrial uses.

Water Rights

A fundamental component of agricultural development in the West has been the establishment of a complex system of water rights. A series of acts in 1866, 1870, and 1877 by the federal government granted individual states the power to distribute water among users (Frederick and Hansen). Since that time the federal government has declined to provide a national policy on ground and surface water allocation. The result is a complicated legal system, each state having its own network of laws governing the existence, acquisition, and exercise of water rights.

Quite simply, a water right accords the user to take, possess, control, and use water. The right of use is a property right protected by government jurisdiction and the system which recognizes it. In essence then, a water right is a publicly protected entitlement to use water in some manner and degree (Weatherford). Irrigation, stock watering, and domestic use account for nearly all water used by the agricultural production sector.

Two basic doctrines govern the use of surface and groundwater in the West. The riparian doctrine accords to the owner of land contiguous to a watercourse a right to the use of water on such land. This doctrine, adopted as part of the common law of England, is the dominant water distribution law in the eastern United States (Hutchins, 1971). The holder of a riparian right can make any use of water as long as he/she uses it on riparian land in what is termed "reasonable use." Under the reasonable use rule, each riparian proprietor may use the water for any beneficial purpose, provided the intended use is reasonable with respect to the needs of other proprietors on the system. Under a riparian system, there exists no priority in right among users; in the case of water shortage, all users share in reducing consumption.

The riparian doctrine proved inadequate for development of lands not adjacent to watercourses. The law of priori appropriation grew out of the need for greater flexibility in the legal framework to encourage investment in water resources. Settlers had little incentive to develop irrigation systems if the legal framework allowed future upstream development priority use of the water supply. Consequently, the western states developed the principle of "first in time, first in right." The doctrine of prior appropriation accords a person who first develops and uses a water supply, the right to continue this use in preference to those who come later (PNW River Basin Commission). An important exception to this rule is that in most states certain preferred uses such as municipal and domestic uses receive their full appropriation regardless of priority.

The appropriation doctrine is unique in that rights are not necessarily tied to ownership of land or use of water on land bordering the water source. The appropriative acquired right is a real property right; it can be sold, transferred, mortgaged, or bequeathed, but frequently only if attached to the land for which it was obtained. The appropriative right pertains to a specific use, is fixed to a specific location of use, and the amount of time of the diversion is usually specified. For this reason, the law of prior appropriation facilitated western expansion by providing a framework for the parcelling of water to a large number of irrigators.

For much of the past century, conflict between the two doctrines has existed in a majority of the western states (Hutchins, 1974). The typical response has been some of modification of the riparian principle. Consequently, many states differ in their degree of recognition of the two rights. The states regarded as the "arid states" (Idaho, Montana, Wyoming, Nevada, Utah, Arizona, New Mexico, and Colorado) have eliminated the riparian doctrine on the grounds that it was unsuited to conditions of their region. The remaining states operate under a "land system" in which both doctrines are recognized. California, however, is the only state where the riparian doctrine has much significance (Ostrom).

As the water resource has become more constraining, the doctrine of prior appropriations has become increasingly criticized for encouraging misappropriation of available water. The primary defect of the law is that the doctrine is a rule of capture. Anyone anticipating an increase in future water demand and higher prices of water can exploit that insight only by investing in diversion works (Williams). Such projects are often premature and thus economically unjustifiable under an efficiency criterion. To the extent that uneconomical investment occurs, the beneficial use requirement can lead to a misallocation of the available water resource.

Despite the intent to prevent the wasteful use of water resources, "beneficial use" provisions often inhibit the producer's incentive to conserve. Water saved through on-farm improvements that increase application efficiency frequently cannot be used on additional land due to the appurtenancy requirement. Therefore, the water saved may be declared as surplus since it does not have a beneficial use on the land it is appurtenant to. In this manner, the "beneficial use" doctrine can discourage conservation or more efficient use of water at the farm level. Initially, the doctrine of absolute ownership was virtually the sole governor of groundwater allocation. This system, borrowed from English common law, entitles an overlying land owner to withdraw groundwater without liability to any other person using the groundwater source (Weatherford). Water use decisions were constrained solely by the profitability of pumping.

Only recently has the allocation of groundwater rights come to the forefront as a controversy. Groundwater use resulting in increased pump lifts, declining well yields, salt water intrusion and jeopardization of some surface water rights has resulted in some conflict. Consequently, a number of doctrines governing the use of groundwater have superceded the early doctrine of absolute ownership.

The allocation of underground streams, is generally governed by the laws of surface watercourses. Most western states have adopted the concept of prior appropriation to groundwater as long as the water is used in a beneficial manner and does not affect prior appropriations. Varying degrees of public control and administration are asserted over groundwater under this system. The American rule of reasonable use and the doctrine of correlative rights are two important doctrines governing groundwater allocation in recent years

Despite the recent proliferation of regulations on aquifer pumping, farmer costs continue to be the principle regulator of groundwater use. These costs did not provide much of a deterrent to pumping until the 1970s. Recently, rising energy costs coupled with lowering water tables have encouraged the use of water and energy-saving innovations. Nevertheless, it does not follow that groundwater or surface water sources are used efficiently. Institutional constraints on water transfer, externalities, and laws regulating aquifer use continue to cause inefficiencies in the allocation of water resources.

The ability of policymakers to modify the current system of water rights is important if scarce water resources are to be allocated to or preserved for higher valued uses. Transfers of water among alternative users are subject to the criterion of non-injury to other existing water right holders. This criterion is particularly difficult to meet when transferring water involves a change in consumptive use, place of use, or point of diversion. This criterion may have to be relaxed if gains in efficient allocations of water are to be achieved.

Perhaps the most difficult step in this reallocation process will be the clear definition of existing agricultural water rights. It is apparent that our present system of laws and institutions must be made more definitive with respect to what is actually implied by the ownership of a water right. The adoption of any reallocation system must first be preceded by judicial and administrative proceedings aimed at quantifying existing rights.

Irrigation Organizations

The type of organization employed to deliver irrigation water to the farm gate affects the efficiency of agricultural water use. Current and expected future economic conditions are quite different from those that prevailed when existing distribution organizations were established. In 1978 about 75 percent of the total water used for irrigation was supplied by some form of organization; the remaining 25 percent was obtained from on-farm sources (Census of Agriculture). Consequently, if conservation, factor substitution, and transfer of water to higher-valued uses are to take place, distribution organizations must be integrally involved.

Three forms of irrigation water delivery organizations dominate western agriculture. Unincorporated mutual water companies, incorporated mutuals, and irrigation districts account for 95 percent of all water organizations and 92.6 percent of all acreage irrigated by organizations (Census of Agriculture).

A mutual company is a non-profit body of irrigators voluntarily organized for the purpose of supplying water to its members. An incorporated mutual is merely a larger, more formal version of the unincorporated mutual. This organization type accounts for about one-third of the number and acreage served by irrigation organizations. Irrigation districts are the most important form of irrigation service organization, transmitting water to half of the total acreage served by irrigation organizations. This organizational form combines many of the features of mutual water companies with the governmental powers of taxation and eminent domain. The truly unique feature of the water district is that it is obligated to provide a service to a given producer that is equivalent to the service provided other producers in the same user class. Therefore, the district has a public utility type obligation that is not found in other organization forms.

An inter-agency task force responsible for assessing the potential of irrigation efficiency concluded that reducing conveyance losses be the foremost objective in agricultural water conservation policy. Reductions in conveyance losses resulting from lining canals and piping, and realignment or enlargement of canals and control structures is estimated at 3.1 million acre-feet per year for western farms (Frederick). The institutional organizations that govern, manage, and operate these water delivery systems must be closely involved in plans to improve the efficiency of water use.

Water Service Agencies

If the problem of water scarcity is to be adequately addressed in the coming years, the public institutional structure responsible for the development, supply, and management of the water resource will surely be impacted. As surface water supplies have developed, a complex network of enterprises and agencies (both public and private) have evolved. These entities interrelate in a manner similar to a well-developed industry, many agencies or firms specialize in a single function, while others become multifunctional. The interrelationship among agencies within the water industry are generally specified in contractual agreements and legislation specifying the means of interagency coordination.

Water is managed in the United States for multiple purposes, including navigation, flood control, municipal, industrial, agricultural, and domestic purposes. In any institutional setting, the relative importance of these programs is dependent upon history, custom, and demand. Generally speaking, governmental agencies instituting water resource policy serve three basic functions: water quantity control, planning and development of water resource utilization, and water quality control (Ostrom).

In general, the USBR and the U.S. Army Corps of Engineers have become the primary large-scale water resource development and management agencies of the federal government. The U.S. Army Corps of Engineers is the primary federal agency involved in navigation on inland waterways and flood control. The coordination of flood control and navigation with alternative water uses has become an integral function of the Corps.

Many Corps projects have become important water storage facilities for agricultural producers, as well as hydroelectric facilities critical to maintaining low electrical energy costs.

An historical review of the USBR's development and enabling legislation was given earlier in this chapter. Today, the USBR's functions are somewhat similar to those specified in the initial legislation. In addition to its planning and construction responsibilities, the USBR produces and markets electricity (Holmes). The provision of water for municipal and domestic purposes has become an increasingly more important function of the USBR; however, providing water to irrigators remains its primary commitment.

When federal agencies undertake large multi-functional water projects, they typically apply to the affected states for a permit to appropriate the water to be involved in the development activity. The water can then be sold by contract to water users, typically represented by an intermediary such as a water district. Through this means, the federal agencies have retained considerable control over water developed from federal projects. As a consequence of this control, the federal agencies responsible for water resource development will not only impact future augmentation of water supplies, but also exert considerable influence on water use and allocation.

The administrative system responsible for coordinating and controlling activities affecting the states' water resources differ dramatically from state to state. Regardless of their structure, these agencies provide the most critical link to attaining more efficient allocation of scarce water resources. State water resource agencies are the obvious choice to administer any type of water banks or markets to facilitate water transfer. Additionally, any change in water rights to encourage conservation would be implemented through this system.

Non-Institutional Factors

The future of irrigated agriculture will undoubtedly be affected by the limited availability of water. A variety of environmental problems associated with irrigation will also impact future water use decisions. Finally, economic factors stemming from high energy prices and competing water uses may affect the viability of irrigated agriculture.

Water Supplies. Throughout the twentieth century expansion of irrigation in the 17 western states has been spurred by augmentation of available water supplies. Total surface withdrawals have fluctuated since the 1950s, but have not continued the previous upward trend (Frederick). Consequently, growth of western irrigation in the last three decades has been based almost entirely on groundwater withdrawals. Opportunities for increasing groundwater withdrawals are becoming increasingly limited due to higher energy costs and declining water tables. In some regions, most notably the Southern Plains, these factors have already begun to curtail irrigation.

Water available for irrigation is not likely to be significantly increased by less conventional sources of supply. Winter cloud seeding will probably not offer much assistance in the immediate future. Also, the costs of water importation and desalination are prohibitively high for providing irrigation water.

Environmental Factors. Although a variety of environmental problems are associated with irrigation, four in particular may have a significant effect on water use. Groundwater depletion, low stream flows, salinity, and pollution from agricultural chemicals may all impose significant constraints on irrigation. An estimated 25 to 35 percent of the West's irrigated lands have salinity problems; however, lands where salinity is likely to curtail production comprise a much smaller percentage (Frederick). Infiltration of agricultural chemicals is not likely to limit agricultural production, but may impede water-use efficiency. On-farm management efforts to keep nitrate levels in ground and surface water in reasonable limits can also be accompanied by increases in on-farm application efficiency.

Economic Factors

The combination of rising energy costs and diminishing water supplies may have serious consequences on future irrigation development and water use. Groundwater users are not the only producers impacted by this phenomenon.

Many producers utilizing surface sources must pump water great distances and heights. The growth of sprinkler irrigation to improve irrigation efficiency and sometimes to curb problems of labor management, soil erosion, or water pollution is also creating an increased vulnerability to the rising costs of energy.

Farmers do have a wide range of opportunities available for responding to high energy costs and reduced water supplies. Irrigation scheduling, tail-water reuse, and improved pumping plant and application efficiency already are economically justifiable under a wide range of conditions. Future innovations designed to attain even higher energy and water efficiencies will become available as the economic conditions facing irrigators change over time. The remainder of this book addresses a number of these opportunities.

REFERENCES

Boris, Constance M. and John V. Krutilla. "Water Rights and Energy Development in the Yellowstone River Basin." Resources for the Future, Baltimore, The Johns Hopkins Press, 1980.

Claiborn, Brent A. "Predicting Attainable Irrigation Efficiencies in the Upper Snake River Region." Idaho Water Resource Research Institute, May 1975.

Frederick, Kenneth D. "The Future Role of Western Irrigation." The Southwestern Review of Management of Economics, 1:19-33, Spring 1981.

Frederick, Kenneth D. and James C. Hansen. "Water for Western Agriculture." Resources for the Future, Baltimore, The Johns Hopkins Press, 1982.

Golze, Alfred. "Reclamation in the United States." Caldwell, Idaho, Caxton Printers, 1961.

Holmes, Beatrice H. "History of Federal Water Resources Programs and Policies, 1961-70." U.S. Department of Agriculture; Economics, Statistics, and Cooperative Service, Misc., Publ. #1379, September 1979.

Hutchins, Wells A. "Water Right Laws in the Nineteen Western States." 3 Volumes, U.S. Department of Agriculture, Economic Research Service, Washington D.C., 1971, 1974, 1977.

Israelson, Orson W. and Vaughn E. Hansen. "Irrigation Principals and Practices." New York, John Wiley and Sons, 1962.

Jensen, Marvin E. "Overview--Irrigation in U.S. Arid and Semiarid Lands." Water-Related Technologies for

Sustaining Agriculture in U.S. Arid and Semiarid Lands, Office of Technologic Assessment, October 1982.

Ostrom, Vincent. "Institutional Arrangement for Water Resource Development." National Technical Information Service, 1971.

Pacific Northwest River Basin Commission. "Columbia-North Pacific Region Comprehensive Framework Study: Legal and Administrative Background." Appendix III, March 1970.

Sloggett, Gordon R. "Energy and U.S. Agriculture: Irrigation Pumping 1974-80." AER 495, U.S. Department of Agriculture, Economic Research Service, December 1982.

Trelease, Frank J. "Water Law, Policies, and Politics: Institutions for Decision Making." In _Western Water Resources, Coming Problems and the Policy Alternatives_, The Federal Reserve Book of Kansas City; Boulder, Colorado, Westview Press, 1980.

United States Department of Agriculture, Soil Conservation Service. "Basic Statistics--1977 National Resource Inventory (NRI)." February 1980.

United States Department of Commerce, Bureau of Census. "1978 Census of Agriculture, Vol. 4, Irrigation." 1981.

Weatherford, Gary. "Acquiring Water for Energy: Institutional Aspects." Water Resources Publications, Littleton, Colorado, 1982.

Williams, Stephen F. "The Requirement of Beneficial Use or a Cause of Waste Water Resource Development." _Natural Resources Journal_, 23:7-23, January 1983.

3
Energy and Water Management with On-Farm Irrigation Systems

Edwin B. Roberts, Richard H. Cuenca, and Robert M. Hagan

INTRODUCTION

A wide variety of irrigation system types are available. Considerable energy could be saved by selection of the most energy-efficient system for each site. However, farmers generally select systems according to economic efficiency, not energy or water efficiency. Minimum cost and minimum energy use do not necessarily coincide. In some cases the minimum irrigation cost is obtained by using relatively energy-intensive systems in order to reduce the use of labor, water, and other resources. Water use efficiency, energy use efficiency, and economic efficiency as well as the principles of optimization of irrigation systems are discussed in other chapters.

On many farms it may be too expensive to use the irrigation systems with the absolute minimum energy requirements, but there are still many opportunities for reducing irrigation energy requirements. Because energy prices have been rising much faster than other prices in many parts of the West in recent years, decisions made when existing systems were installed are no longer valid in some cases. Also, system design procedures and irrigation management practices have been relatively crude in some cases. Modification or replacement of existing irrigation systems, along with improved management, could both save energy and reduce costs on some farms.

The purpose of this chapter is to review the most common types of on-farm irrigation systems and discuss some of the problems in selecting an improved system for a specific farm. Energy requirements for irrigation, both direct and embodied energy, are discussed and the use of a computer model to calculate energy requirements is described.

IRRIGATION APPLICATION EFFICIENCY

Most of the energy used on-farm for irrigation is used for pumping water. Therefore, one of the main ways to conserve energy is to conserve water. Irrigation application efficiency is a measure of the relative effectiveness of different irrigation systems in conserving water.

One commonly used definition of irrigation application efficiency is the ratio of the amount of water stored in the crop root zone through irrigation to the amount of water applied. The water not stored in the root zone may be lost (depending on the type of system) through evaporation to the air, deep percolation below the crop root zone, runoff from the end of the field, and seepage and evaporation from distribution ditches. Some of the losses result from the physical processes involved in irrigation and are unavoidable. Also, much of the water lost through runoff and deep percolation returns to rivers or groundwater storage for reuse at another time or location.

In some cases water is actually wasted through excessive applications resulting from poor management or improper system design. Eliminating excessive applications will not only save water but also save energy and reduce irrigation costs.

Where irrigation management is already good, further improvements in irrigation efficiency will require changes in the irrigation system. These changes involve trade-offs between water requirements, energy requirements, and costs. For example, replacing a surface irrigation system with a sprinkler system may improve the irrigation efficiency and reduce water use, but the capital investment in the sprinkler system will be high and the energy required will very likely increase. Results of such changes will vary from farm to farm. There is no best system for all cases.

It is difficult to compare types of irrigation systems on the basis of irrigation efficiency figures reported in the literature because irrigation efficiency is related to the uniformity and adequacy of application and measures of these factors are seldom reported. Irrigation applications are never completely uniform over a field. Some areas of a field receive more than the desired application and some receive less. It is easier to achieve high irrigation efficiency figures with low irrigation adequacy, but poor uniformity of application may then reduce crop yields and quality.

One common rule of thumb is to apply irrigations so that the one-fourth of the field receiving the lightest application receives an average application equal to the desired depth of application. Approximately 87.5 percent of the field will then be at least adequately irrigated and 12.5 percent will be slightly under-irrigated. It is generally believed that this amount of under-irrigation will not significantly reduce yields, especially because the part of the field being under-irrigated may change from one irrigation to the next. If the level of adequacy is higher, yields may be reduced by serious over-irrigation in some parts of the field, leading to waterlogging of the soil and excess leaching of plant nutrients out of the crop root zone. More complete discussions of irrigation efficiency, adequacy, and uniformity are in Chapters 4 and 5.

TYPES OF IRRIGATION SYSTEMS

This section describes most of the commonly used types of irrigation application systems and presents some generally accepted values for irrigation application efficiency, all based on an irrigation adequacy of 87.5 percent. More complete descriptions of the available types of systems are given by Addink, et al.; American Association for Vocational Instructional Materials; Hart, et al.; and Pair.

Water requirements, energy requirements, labor requirements, and costs for the major types of systems are shown in Tables 3.1 and 3.2. Comparison of the resource requirements and costs for different systems is difficult because they are affected by field size and shape, soil texture, slope, crop types, weather conditions, water source, wage rates, energy rates, and other factors, and each type of system is affected somewhat differently by changes in conditions. Therefore Table 3.1 shows ranges of values which take into account the most favorable and least favorable conditions normally encountered in the West. All of the values are for well-designed and well-managed systems. Table 3.2 compares the resource requirements and costs for the different system types for one set of conditions which we believe represents fairly "typical" or "average" conditions in the West.

Table 3.1

Range of water, energy, labor requirements and costs for good irrigation systems[a]

System Code[b]	PAE[c] (%)	Annual Water Requirements (AF/acre/ft)[d]	Annual Energy Requirements (kwh/acre/ft)[e]: Application	Annual Energy Requirements (kwh/acre/ft)[e]: Lift from Well (per ft lift)[f]	Annual Labor Requirements (hr/acre/ft)[g]	Capital Cost of Application System[h] ($/acre)	Total Annual Cost[i] ($/acre): 1.5 ft Net Application, 50-ft Lift	1.5 ft Net Application, 500-ft Lift	4-ft Net Application, 50-ft Lift	4-ft Net Application, 500-ft Lift
B	50–80[j]	1.56–2.50[h]	0	2.13–3.41	1–6[l]	100–600[m]	47–207[n]	96–527[n]	66–387[n]	199–1,238[n]
BP	50–80[j]	1.25–2.00[h]	3.41–27.3[p]	1.71–2.73	1–6[l]	175–1,000[q]	55–252[n]	94–580[n]	70–415[n]	162–998[n]
BT	60–90[j]	1.39–2.20[h]	10.40–18.4[r]	1.90–3.00	1–6[l]	150–650[s]	53–208[n]	91–448[n]	72–388[n]	190–1,127[n]
BPT	60–90[j]	1.08–1.70[h]	13.30–41.6[t]	1.47–2.32	1–6[l]	225–1,050[u]	61–257[n]	96–474[n]	78–426[n]	170–1,005[n]
LB	70–95	1.32–1.79[h]	0	1.80–2.44	.5–3[v]	100–600[m]	43–161[n]	85–390[n]	55–264[n]	168–873[n]
LBP	70–95	1.05–1.43[h]	00.0–19.5[w]	1.43–1.95	.5–3[v]	100–850[x]	42–189[n]	75–372[n]	53–288[n]	142–775[n]
F	75	1.67[h]	0	2.28	2–12[y]	100–600[m]	58–328[z]	111–541[z]	88–607[z]	230–1,176[z]
FP	75	1.33[h]	9.10–45.4[aa]	1.82	1.5–9[bb]	185–900[cc]	64–329[z]	107–498[z]	88–551[z]	201–1,004[z]
FT	90	1.53[h]	33.80[r]	2.09	2–12[y]	150–650[s]	65–335[z]	114–531[z]	95–616	226–1,137[z]
FPT	90	1.12	47.1–100[dd]	1.53	1.5–9[bb]	235–950[ee]	71–339[z]	92–421[z]	96–567[z]	191–948[z]
H	70–80	1.25–1.43	137–264[ff]	1.71–1.95	1–3[gg]	125–425	36–162	76–344	62–314	168–801
W	70–80	1.25–1.43	222–273[hh]	1.71–1.95	.5–1.5[ii]	250–550	57–167	96–350	84–296	191–783
SS	70–80	1.25–1.43	137–264[ff]	1.71–1.95	1/yr[jj]	800–1,200	196–250	236–432	213–346	319–833
P	70–80	1.25–1.43	128–283[kk]	1.71–1.95	1/yr[kk]	550–600	111–172	151–355	217–273	233–760
D	90	1.11	83[ll]	1.52	4/yr[mm]	825	213–251	249–393	226–296	320–674

Footnotes for Table 5.1 on next page.

Footnotes - Table 3.1

[a]For well-designed and well-managed irrigation systems. Based on previously published information, costs obtained from equipment dealers and manufacturers, and unpublished calculations by the authors.

[b]B = graded border with ditches; BP = graded border with buried plastic pipeline system; BT = graded border with ditches and tailwater reuse; BPT = graded border with pipeline and tailwater reuse; LB = level border; LBP = level border with buried plastic pipeline; F = graded furrow with ditches; FP = graded furrow with buried plastic mainline and gated aluminum laterals; FT = graded furrow with ditches and tailwater reuse; FPT = graded furrow with buried plastic mainline, gated aluminum laterals, and tailwater reuse; H = hand-move sprinkler; W = well-line sprinkler; SS = solid-set (permanent) sprinklers; P = center-pivot; D = drip.

[c]Potential application efficiency.

[d]Acre-feet/acre per foot of net water application to the root zone.

[e]kwh/acre per foot of net water application, assuming a pumping plant efficiency of 75 percent (attainable by the best designs of new pumps).

[f]Per foot of lift from water surface in well to pump discharge.

[g]Man-hours/acre per foot of net water application.

[h]Not including pumping plant and well.

[i]Includes depreciation on application system and pumping plant but not well (which is assumed to be the same for all irrigation systems), interest (10 percent net inflation-free interest rate is assumed), taxes and insurance (assumed to be equal to 2 percent of capital cost), repairs, irrigation labor (based on wage rate, including benefits, of $4-$8/hour), and energy (based on electric rates of $.02-$.08/kwh).

[j]Depends on soil texture, slope, and net depth of application.

[k]Includes assumed 20 percent seepage and evaporation loss from ditches.

[l]Based on labor requirement of 0.5-1 hour/acre/irrigation and net application of 2-6 inches per irrigation.

[m]Land grading cost.

[n]Includes $50/acre every 5 years to "touch-up" land grading and $15/acre/year to install and take out borders.

[p]Based on pressure head of 2-10 feet.

[q]Based on land grading cost of \$100-\$600/acre and pipeline cost of \$75-\$400/acre (depending upon field size, border length, and flow rate).

[r]For tailwater reuse pumping.

[s]Includes \$100-\$600/acre for land grading and \$50/acre for tailwater reuse system.

[t]Based on pressure head of 2-10 feet. Also includes energy for tailwater reuse pumping.

[u]Includes \$100-\$600/acre for land grading, \$75-\$400/acre for pipeline, and \$50/acre for tailwater reuse system.

[v]Based on labor requirement of 0.25-0.5 hour/acre/irrigation and net application of 2-6 inches/irrigation.

[w]Based on pressure head of 0-10 feet (with some very large level borders the pump discharges into the border with only a very short pipe).

[x]Includes \$100-\$600/acre for land grading and \$0-\$250/acre for pipeline.

[y]Based on labor requirement of 1-2 hour/acre/irrigation and net application of 2-6 inches/irrigation.

[z]Includes \$50/acre every 5 years to "touch-up" land grading and \$20-\$80/acre (depending upon type of crop) to install and take out furrows.

[aa]Based on pressure head of 5-25 feet.

[bb]Assumes gated pipe reduces labor requirements 25 percent compared with ditches and siphons.

[cc]Includes land grading cost of \$100-\$600/acre and pipeline cost of \$85-\$300/acre.

[dd]Based on pressure head of 5-25 feet. Also includes energy for tailwater reuse pumping.

[ee]Includes \$100-\$600/acre for land grading, \$85-\$300/acre for pipe system, and \$50/acre for tailwater reuse system.

[ff]Based on sprinkler pressure of 25-45 psi (pressure head at the pump of 80-135 feet).

[gg]Based on labor requirement of 0.5 hour/acre/irrigation and net application of 2-6 inches/irrigation.

[hh]Based on sprinkler pressure of 45 psi (pressure head at the pump of 130-140 feet).

[ii]Based on labor requirement of 0.25 hour/acre/irrigation and net application of 2-6 inches/irrigation.

[jj]One (1) hour/acre/year.

[kk]Based on end pressure of 20-40 psi (pressure head at the pump of 75-145 feet).

[ll]Based on design pressure of 15 psi (pressure head at the pump of 55 feet).

[mm]Four (4) hours/acre/year.

Table 3.2
Typical water, energy, labor requirements and costs for good irrigation systems

System Code	PAE %	Annual Water Requirements (AF/acre/ft)	Annual Energy Requirements (kwh/acre/ft)		Annual Labor Requirements (hr/acre/ft)	Capital Cost of Application System ($/acre)	Total Annual Cost[a] ($/acre)			
							1.5 ft Net Application		4 ft Net Application	
			Application	Lift from Well (per ft lift)			50-ft Lift	500-ft Lift	50-ft Lift	500-ft Lift
B	70	1.79[b]	0	2.44	1.5[c]	150[4]	69[e]	212[e]	118[e]	499[e]
BP	70	1.43	13.7[f]	1.95	1.5[c]	275[7]	82[e]	197[e]	123[e]	384[e]
BT	85	1.56[b]	13.8[h]	2.13	1.5[c]	200[9]	73[e]	180[e]	123[e]	455[e]
BPT	85	1.21	25.4[j]	1.65	1.5[c]	325[11]	88[e]	185[e]	132[e]	390[e]
LB	90	1.39[b]	0	1.90	.75[l]	300[13]	74[e]	185[e]	106[e]	402[e]
LBP	90	1.11	15.2[n]	1.52	.75[l]	400[15]	88[e]	177[e]	118[e]	354[e]
F	75	1.67[b]	0	2.28	3.0[q]	150[4]	87[r]	220[r]	157[r]	512[r]
FP	75	1.33	45.4[s]	1.82	2.25[t]	300[20]	101[r]	207[r]	162[r]	445[r]
FT	90	1.53[b]	33.8[h]	2.09	3.0[q]	200[9]	94[r]	216[r]	165[r]	490[r]
FPT	90	1.12	100.0[v]	1.53	2.25[t]	350	110[r]	162[r]	175[r]	413[r]
H	80	1.25	171.0[x]	1.71	1.5[y]	225	79	179	145	411
W	80	1.25	239.0[z]	1.71	.75[aa]	350	100	200	166	432
SS	80	1.25	171.0[x]	1.71	1/yr[bb]	1,200	216	315	260	526
P	80	1.25	137.0[cc]	1.71	1/yr[bb]	600	138	237	177	443
D	90	1.11	83.0[dd]	1.52	4/yr[ee]	825	231	320	216	497

Footnotes for Table 2 on next page.

Footnotes for Table 3.2

(See also the footnotes for Table 3.1)

[a]Assumes wage rate, including benefits, of $6/hour and energy rate of $.05/kwh.

[b]Includes assumed 20 percent seepage and evaporation loss from ditches.

[c]Based on labor requirement of 0.5 hour/acre/irrigation and net application of 4 inches/irrigation.

[d]Land-grading cost.

[e]Includes $50/acre every 5 years to "touch-up" land grading and $15/acre/year to install and take out borders.

[f]Based on pressure head of 7 feet.

[g]Includes land grading cost of $150/acre and pipeline cost of $125/acre.

[h]For tailwater reuse pumping.

[i]Includes land grading cost of $150/acre and tailwater reuse system cost of $50/acre.

[j]Based on pressure head of 7 feet. Also includes energy for tailwater reuse pumping.

[k]Includes land grading cost of $150/acre, pipeline cost of $125/acre, and tailwater reuse system cost of $50/acre.

[l]Based on labor requirement of 0.25 hour/acre/irrigation and net application of 4 inches/irrigation.

[m]Includes land grading cost of $300/acre (assuming level borders require more precise grading than graded borders).

[n]Based on pressure head of 10 feet.

[p]Includes land grading cost of $300/acre and pipeline cost of $100/acre.

[q]Based on labor requirement of 1 hour/acre/irrigation and net application of 4 inches/irrigation.

[r]Includes $50/acre every 5 years to "touch-up" land grading and $20/acre to install and take out borders.

[s]Based on pressure head of 25 feet.

[t]Assumes gated pipe reduces labor requirements 25 percent compared with ditches and siphons.

[u]Includes $150/acre for land grading and $150/acre for pipeline system.

[v]Based on pressure head of 25 feet. Also includes energy for tailwater reuse pumping.

[w]Includes $150/acre for land grading, $150/acre for pipeline system, and $50/acre for tailwater reuse system.

[x]Based on sprinkler pressure of 35 psi (pressure head at the pump of 100 feet).

[y]Based on labor requirement of 0.5 hour/acre/ irrigation and net application of 4 inches/irrigation.

[z]Based on sprinkler pressure of 45 psi (pressure head at the pump of 140 feet).

[aa]Based on labor requirement of 0.25 hour/acre/ irrigation and net application of 4 inches/irrigation.

[bb]One (1) hour/acre/year.

[cc]Based on end pressure of 20 psi (pressure head at the pump of 80 feet).

[dd]Based on design pressure of 15 psi (pressure head at the pump of 55 feet).

[ee]Four (4) hour/acre/year.

Graded Border

In graded border irrigation (also known as border-check, border-strip, strip-check, or simply as border or flood irrigation) the field to be irrigated is divided into sloping strips which are separated by parallel levees or ridges. Water is turned in at the upper end of the border and flows down-slope as a sheet, guided by the ridges. Graded border irrigation is suited to close-growing field crops such as alfalfa, pasture, and small grains, and also to orchards and vineyards. It is suited to soils with moderately-low to moderately-high water intake rates. Slopes can be as little as 0.1 percent or as much as 2.0 percent (4.0 percent with sod crops).

Proper design of graded borders involves balancing the inflow rate and time with the length and slope of the border and the soil characteristics so that the moving sheet of water covers each part of the border for approximately the same time and water has an equal opportunity for infiltration at each point. Uniform soils and slopes are essential for good irrigation efficiency. Border lengths may vary from about 300 to 2,600 feet. Lengths of one-fourth mile are probably most common. Widths may vary from about 10-20 feet and are chosen partly for efficient operation of cultural and harvesting equipment.

Merriam defines the potential application efficiency (PAE) as the water application efficiency that should be obtained by an irrigation system when it is properly used. The PAE for graded border irrigation (without tailwater reuse) varies from about 50 percent to 80 percent (U.S.

Soil Conservation Service). The higher figure is possible with medium-textured soil, shallow slopes and relatively large net applications of 4-5 inches. Very coarse or very fine textured soil, steeper slopes, and lighter applications reduce PAE. Merriam suggests that the actual application efficiency (AAE), the application efficiency that occurs in practice, is commonly 65 percent or below.

In the simplest variation of graded border irrigation, a head ditch is dug across the top of the field to distribute water to the border-strips. Water moves by gravity flow from the head ditch into and along the border-strip. Thus, no pumping energy is required to operate the system, although pumping may be required to supply water to the head ditch from wells or surface water sources. Unlined earth head ditches and other farm ditches used to convey irrigation water lose water by evaporation, seepage, and spillage. The conveyance efficiency for these ditches is 60-80 percent (Fereres, et al.).

For the graded border system with ditches (System B) in Table 3.1, an 80 percent conveyance efficiency is assumed. Therefore, for each foot of net water application to the root zone, the gross water requirement is 1.56-2.50 feet. With the assumed 78 percent pumping plant efficiency, the energy requirement for pumping this amount of water is 2.13-3.41 kwh per foot of lift. (Energy requirements will be higher with older, less-efficient pumping plants.) The labor requirement for applying the same amount of water typically varies from about 1-6 hours/acre, depending upon border length and the number of irrigations needed to apply 1 foot net to the root zone.

The main investment cost for graded border irrigation is for grading the field to a uniform slope. This typically runs from $100-$600 per acre using the latest laser leveling technology (Daubert and Ayer).

Total annual costs are shown in Table 3.1 for relatively light to relatively heavy net annual water applications of 1.5-4.0 feet and relatively shallow to relatively deep wells with pumping lifts of 50-500 feet. For graded border irrigation using ditches the range of annual costs is very wide, from $47-$1,238 per acre. Even with the light application and low pumping lift the range of costs is from $47-$207/acre. However, the extreme low cost is based on the most favorable irrigation efficiency, labor requirement, wage rate, land grading cost, and energy rate while the extreme high annual cost is based on the least favorable conditions. The majority of irrigation systems will be affected by a combination of favorable and unfavorable conditions and the resultant annual costs will fall somewhere in the middle of the range. Table 3.2 is based on moderate or "typical" conditions. The total annual cost for graded borders with ditches varies from $69 to $499 per acre, depending on net annual water application and pumping lift. The following sections on pipeline water distribution and tailwater reuse are related to the use of border irrigation systems. However, their effects can be equally applied to all gravity flow irrigation systems; these include levee, furrow, and corrugation systems.

Pipeline Water Distribution. Pipelines almost completely eliminate conveyance losses. Buried concrete pipelines have been used with graded border irrigation systems for many years, especially for permanent crops such as orchards, vineyards, and permanent pasture. Valves located along the pipelines can be opened to release water at the top end of individual borders or groups of borders. Buried plastic pipelines are now becoming popular. Aluminum or plastic gated pipe laid along the surface can also be used for water distribution in graded border irrigation.

A slight pressure is required in a pipeline to cause water to flow through gates or valves and overcome friction losses. However, less water must be pumped from the original source than in a system using ditches because conveyance losses are eliminated. Usually, the energy required to develop pressure in the pipeline is more than offset by the reduced volume of water pumped whenever the water must be lifted more than 25-50 feet from a well or surface source.

The capital cost of buried plastic pipeline systems for graded borders ranges from about $75-$400/acre, depending on field size and shape, length of borders, and flow rate. As a result of the pipeline cost, the range of annual costs for graded border irrigation systems with pipelines (System BP) shown in Table 3.1 is somewhat higher at the 50-foot lift than for graded borders with ditches (System B). However, at the 500-foot lift the range of costs for System BP is lower.

Tailwater Reuse. A tailwater reuse system is made up of a sump or reservoir to capture the tailwater (runoff) from the lower ends of the borders, a pump, and pipeline to return the water to the top of the field or another field for reuse. A tailwater reuse system can increase the irrigation efficiency (PAE) of a graded border irrigation system to as high as 90 percent. The reused water replaces water otherwise obtained from the original surface or groundwater source.

The total pumping head for a tailwater pump is typically about 30 feet including both the lift from the sump to the top of the field and friction head in the pipeline. Thus, a tail-water reuse system reduces overall irrigation energy requirements wherever the pumping head for the original water supply is more than about 30 feet.

As shown in Tables 3.1 and 3.2, adding tailwater reuse to a graded border/ditch system tends to slightly increase total annual costs at a pumping lift of 50 feet but reduce annual costs at the 500-foot lift (comparing figures for System BT with those for System B).

Level Border

Level border irrigation is also known as basin, level-basin, or dead-level irrigation. Level borders differ from graded borders in that the land between the ridges is level rather than sloping and the ends of the borders are closed so that irrigation water is ponded until it infiltrates into the soil. Usually, a relatively large irrigation stream is used so that the entire strip is covered in a relatively short time. The water is turned off when the desired volume has been applied. For many years some orchards have been irrigated using small basins around individual trees. This system uses small streams of water and has relatively high labor requirements. The development of laser-leveling equipment has made precise leveling of much larger level borders possible. Some level borders are as large as 40 acres. The method is being used extensively in the Southwest (Erie and Dedrick).

Level border irrigation can be used on almost any crop but is most commonly used with close-growing field crops such as alfalfa, pasture, and small grains, and in orchards and vineyards. It is best suited to uniform soils of moderate to low intake rate. Because no water is lost by runoff, PAE can be very high, up to about 70-95 percent, depending upon soil type and rooting depth. AAE tends to be 80 percent or lower (Merriam).

If earth ditches are used and water is available at the edge of the field, no energy is required for application and there is no capital cost other than for land leveling. The labor cost for level borders tends to be about half that for graded borders (American Association for Vocational Instructional Materials).

Because level borders may require more precise land grading than graded borders, we assumed higher land-grading costs for level borders (System LB) in Table 3.2. As a result the total annual cost for the 1.5-foot net application and 50-foot lift is slightly higher for the level border system at $74/acre than for the graded border system at $69/acre. However, with either the greater net application or greater lift, the level border system has a lower annual cost.

Just as with graded border irrigation, the earth ditches used with level borders can be replaced with pipelines. In favorable conditions, level border/pipeline systems (LBP) can have lower water and energy requirements than any other type of system (see Table 3.1). Total annual costs also tend to be lower than for other systems where both annual water applications and pumping lifts are high.

Contour Levee

Contour levee irrigation is similar to level border irrigation except that it is adapted to sloping land. The levees or ridges follow the contours of the land. In contrast to level border irrigation where the border area is level and all water percolates into the soil, the contour levee area is sloped from one levee to the next and excess water is drained off after an irrigation. Contour levees are widely used for growing rice, in which case water is circulated through the basins throughout most of the season. Contour levees are adapted to low intake rate soils which are difficult to irrigate by other surface methods (Hart, et al.).

Graded Furrow

In graded furrow irrigation small, evenly spaced, shallow channels are installed down the slope of the field to be irrigated. The crop is grown in beds between the furrows. Water is applied to the high end of the field and conveyed along the furrows from which it penetrates laterally and downward to wet the root zone of the crop. Graded furrows are best suited to clean tilled crops planted in rows (Hart, et al.).

The method is suited to medium to moderately fine textured soils of relatively high available water holding capacity and conductivities which allow significant movement of water in both the horizontal and vertical directions (Hart, et al.). For best application efficiency, soils and slopes should be uniform and slopes should not exceed about 1.0 percent. The lateral spacing of furrows is usally 2.5 to 3.5 feet depending upon crop and soil characteristics. Furrow lengths range from about 200 feet to 2,640 feet, but 1,320 feet is probably most common.

The PAE for graded furrow irrigation is around 75 percent and the AAE is typically 45-60 percent (Merriam). As with graded borders and level borders, no energy is required for water application if earth or concrete-lined head ditches are used.

One rule of thumb for the labor requirement of graded furrow irrigation is that one irrigator can control a water stream of 2,250 gpm (Fereres, et al.). Another is that furrows of 1,320-foot length require about 1.0 man-hours per acre per irrigation, approximately twice the labor required by graded borders of the same length.

The earth head ditches and siphons commonly used to distribute water to individual furrows can be replaced by gated pipes to eliminate conveyance losses. Gated pipe systems have been reported to reduce irrigation labor requirements by 20-25 percent compared with ditch and siphon systems.

Tailwater reuse systems can be used with graded furrows in the same manner as with graded borders. Tailwater reuse can raise PAE values to about 90 percent.

Primarily because of higher labor requirements, the range of annual costs for graded furrow irrigation systems tends to be higher than for graded border systems, as shown in Tables 3.1 and 3.2. However, the two irrigation methods are generally used on different crops.

Contour Furrow

Contour furrows are the same as graded furrows except that contour furrows are laid out across the steepest slope of the field and are curved to fit the land surface. Operation is the same as with graded furrows.

Corrugation

Corrugations are essentially small, closely spaced graded furrows running parallel to the slope of the field. Corrugation irrigation is usually used with non-cultivated, close-growing crops whereas graded furrow irrigation is used with clean-tilled row crops. Otherwise, operation of the two methods is very similar (Hart, et al.).

Level Furrow

In level furrow irrigation, the land is completely level and the ends of the furrows are blocked to prevent runoff. Water is applied to one end of each furrow at a high enough rate so that the entire length of the furrow will be covered in a relatively short time, the required volume is applied, and then the water is ponded until it infiltrates (Hart, et al.). Often, groups of furrows are enclosed within the ridges of a basin. Operation is then similar to level border irrigation. Such systems are sometimes referred to as "beds-in-basins" or "dead-level furrows."

Hand-Move Sprinkler

A hand-move sprinkler system (sometimes called a portable-set system) consists of one or more lightweight aluminum lateral pipelines connected to a mainline which may be either portable aluminum pipe laid on the surface or buried pipe of steel or plastic. Lateral lines are in sections of 20 to 40 feet in length with quick couplers for easy movement by hand. Each section carries one rotating sprinkler. Successive sets are irrigated by moving laterals to new positions along the mainline. Spacing of lateral positions along the mainline is usually in the range of 20-60 feet. Hand-move sprinklers can be used with almost any crop or soil type. Very tall crops like field corn are uncompatible with hand-move sprinklers because of the difficulty in moving the laterals. Unlike surface irrigation methods, hand-move sprinklers can be used satisfactorily on slopes as steep as 20 percent and with nonuniform soils and slopes.

PAE for hand-move sprinklers is typically about 70-80 percent. AAE is often 60 percent or less because many operators tend to use longer than necessary set times and apply too much water (Merriam).

Labor requirements vary from about 0.35-1.0 man-hours per acre per irrigation depending upon the diameter and length of lateral pipe sections (which determine the weight and the distance the irrigator must walk). Capital costs range from about $125-$425 per acre. Until recently, many hand-move systems used a relatively wide sprinkler spacing of 40 feet by 60 feet and operating pressures of 60-80 psi. The latest sprinkler head and nozzle designs now make it possible to achieve satisfactory uniformity of application with pressures as low as 25 psi in some situations. However, closer sprinkler (and lateral) spacings such as 30 feet by 40 feet or 30 feet by 50 feet are required. The closer sprinkler and lateral spacing does increase the investment costs, however.

Under the most favorable conditions, the total annual costs for hand-move sprinklers (System H) can be lower than for any of the surface irrigation methods, as shown in Table 3.1. However, more typically, the sprinklers tend to be more expensive than some graded border and level border variations where pumping lifts are low (as shown for the 50-foot lifts in Table 3.2).

Hose-Drag Sprinkler

Hose-drag sprinklers are occasionally used in orchards. Mainlines and sub-mains are buried. Flexible hoses, usually carrying from one to three sprinklers are connected to hydrants along the submains. To change sets, irrigators drag the hoses to new positions using the same hydrant connections or new hydrant connections.

Side-Roll Wheel-Move Sprinkler

A side-roll wheel-move sprinkler system has wheels mounted on the lateral line so that the line can be rolled to new positions across the field. Rigid couplers are used so that the entire line can be moved as a unit. Usually a small gasoline engine attached to the middle of the line moves the system. Lateral pipes are usually 4 or 5 inches in diameter and are made of relatively heavy gauge aluminum so that the pipe itself can be used as an axle. Individual sections of lateral pipe are usually 30 or 40 feet in length. Lateral lines are most commonly one-fourth mile long. The laterals are connected to the mainline by flexible hoses at each new position. Wheel diameters range from about 4 to 10 feet.

Some side-roll systems use trailing lines carrying from one to three sprinklers attached by special swivel couplings. The trailing sprinklers make it possible to irrigate a larger area with a given number of side-roll laterals but the trailing lines must be disconnected and moved by hand when a lateral reaches the end of the field and the direction of travel is reversed.

As shown in Tables 3.1 and 3.2, side-roll wheel-move sprinklers (System W) generally have lower labor requirements than hand-move systems but higher capital costs and higher energy requirements. Overall annual costs tend to be slightly higher for the side-roll systems in most situations.

Solid-Set and Permanent Sprinklers

Portable solid-set sprinklers use the same lightweight aluminum laterals as hand move sprinklers but enough lateral lines are used to cover an entire field. Solid-set systems are mainly used to irrigate crops which are cultivated during the irrigation season. The pipes are moved to the field and set up by hand at the beginning of the season and then put back in storage at the end of the season. An entire field may be irrigated in one set or it may be irrigated in sections in which case sets are changed just by switch valves. Labor costs for solid-set systems are much lower than for hand-move systems. However, capital costs are so high, about $1,200/acre, that solid-set systems are only practical with high-value crops in special circumstances. For example, they are commonly used in orchards and vineyards for frost protection in the spring and for crop cooling on excessively hot summer days. Solid-set systems are occasionally used for germination and establishment of vegetable crops, which are then furrow irrigated for most of the irrigation season.

Permanent sprinkler systems are also solid-set systems. However, all main and lateral lines are buried. Sprinklers are mounted on risers which extend above the soil surface. In some orchards and vineyards, very tall riser pipes are used so that the sprinklers are mounted above the tops of the trees or vines. Such systems are called overhead sprinklers. Permanent sprinklers can easily be automated by installation of remote control, electrically operated, valves for changing sets.

Traveling Big-Gun Sprinkler

The traveling big-gun sprinkler system (also known as the single-sprinkler, self-propelled type) uses a very large sprinkler mounted on a wheeled cart and fed by a large diameter flexible rubber hose. The machine is self-propelled and travels across the field moving continuously in a lane guided by a steel cable anchored at the side of the field. One pass across the field irrigates a large rectangular area and then the machine and hose can be moved to an adjacent lane to irrigate another strip. Traveling sprinklers have low labor requirements but operating pressures range from 60-110 psi, higher than for most other types of sprinkler systems.

Center-Pivot Sprinkler

A center-pivot sprinkler system uses a single large diameter lateral line mounted on towers which are A-frames with wheels. The lateral is pivoted from one end so that it irrigates a circular area. Water is supplied to the lateral through the pivot end. Depending on the length of the machine, the area irrigated may be from 20-200 acres or more. The lateral line is usually made of steel, although aluminum is available. Pipe diameters range from about 4.5 to 10 inches. The towers are self-propelled, usually by electric motors mounted in the wheels, and the system is continuously moving during irrigation. A guidance system stops and starts individual towers as necessary to keep the lateral in a straight line. When center pivots are used in square fields, a giant sprinkler mount on the outer end or various types of extending booms may be used to irrigate the corners of the fields.

Since the center-pivot machine moves in a circle, sprinklers toward the outer end cover a larger area than those toward the center. To compensate, larger sprinklers are used near the outer end or else sprinklers are spaced more closely near the outer end.

The older center-pivot systems operated at relatively high pressures, about 60-100 psi at the pivot. Currently because of high energy prices, most systems are being designed to use low-pressure rotating sprinklers with pivot pressures as low as 45 psi or circular spray nozzles with pivot pressures as low as 20 psi. The water application rate at the outer end of the lateral is high and may exceed the rate at which water infiltrates into some soils, resulting in surface runoff a problem that is increased by low pressure systems. Systems using low-pressure spray nozzles have the highest applications rates and are not well suited for use on heavy soils.

Linear Sprinkler

Linear sprinkler systems are also called lateral-move, traveling lateral, or continuously moving straight lateral systems. The lateral line and towers of a linear system are very much like those of a center-pivot system, but, with the linear system, the entire lateral moves continuously across the field during an irrigation. Water may be supplied to a linear machine through an intake pipe or hose which is dragged along the bottom of a concrete lined ditch or through a long flexible rubber hose connected by hand at different points along a mainline. Some of the newest designs can automatically connect to hydrants spaced along buried mainlines. Because all parts of the lateral line travel across the field at the same speed, sprinkler size and spacing are uniform. Linear sprinklers do not have the same problems with high application rates as center-pivot sprinklers and low pressure spray nozzles can be used successfully on most soils.

Drip

Drip, or trickle irrigation, is the frequent, slow application of water to soil through mechanical devices or holes called emitters or drippers (Pair). The emitters are supplied with water by a network of plastic, main, submain, and lateral lines. Each emitter discharges water at a very low rate and the volume of soil wetted by each emitter is relatively small. Therefore, several emitters may be needed to water full-grown trees in orchards. Drip irrigation is most commonly used in orchards and vineyards although it has been tried experimentally with certain row crops. The objective is to supply water almost continuously (daily) to each plant to replace the water lost through evapotranspiration. Potential application efficiency (PAE) with drip irrigation is high (80-90 percent) and operating pressures are generally very low. In some situations yield increases have been reported compared with other irrigation methods. Water savings are especially good in immature orchards because only the soil close to the trees must be watered, not the bare soil between rows. Filtration of the water supply is required to prevent clogging of the small orifices in emitters. Careful hydraulic design is required to achieve good application uniformity. The capital cost of drip systems is relatively high. As shown in Table 3.2, although water, energy, and labor requirements of drip systems tend to be lower than for surface irrigation systems or hand-move sprinklers, total annual costs tend to be considerably higher.

TYPES OF ENERGY USED FOR IRRIGATION

Energy is used in various ways for irrigation, both directly and indirectly. Sprinkler and drip irrigation systems require direct pumping energy to provide the necessary discharge pressure at the nozzles or emitters, overcome friction losses in pipelines, and overcome head losses that occur in various joints, reducers, bends in the pipe, and possibly, filtration units. Some surface irrigation systems operate with no on-farm pumping energy because water is delivered to the farms by irrigation districts either in pressurized pipelines or in canals slightly above field elevations. Water then can move by gravity flow from the irrigation district turnouts into the field head ditches and then into the borders or furrows. In other cases pipelines are used for water distribution instead of head ditches and on-farm pumping is required to overcome friction losses and cause water to flow through the outlets.

Where irrigation water is supplied from an on-farm well, pumping energy is required to lift water to the field surface and overcome friction losses in the column pipe (the pipe that carries water up through the well) and the driveshaft bearings. In many cases the same pump is used to both lift water from a well and pressurize an irrigation system.

Properly designed, new pumps have efficiencies of about 80 percent and the electric motors used for pumping have efficiencies ranging from about 88-93 percent (depending upon size). Therefore the overall efficiency of a good electric pumping plant (pump efficiency x motor efficiency) ranges from about 70-75 percent. Because of wear and installation problems, the average efficiency of pumping plants in use is around 60 percent, but frequently lower. The standard formula (from Knutson, et al., 1981) for calculating electrical energy requirements for pumping is:

$$\text{kwh} = 1.024 \times \frac{\text{(total head)} \times \text{(acre-feet)}}{\text{OPE (decimal)}}$$

Total head is measured in feet. Pressure of one pound-per-square-inch is equal to a head (lift) of 2.31 feet. For an overall plant efficiency (OPE) of 60 percent, the energy requirement for pumping is 1.71 kwh/acre-foot/foot of head. For an OPE of 75 percent, the energy requirement is 1.37 kwh/acre-foot/foot of head.

Much of our electricity is generated in plants using fossil fuel (coal, oil, and natural gas). The typical efficiency for generation and transmission of electricity from these plants is about 31 percent. Thus the overall plant efficiency for electric pumping plants, in terms of fossil-fuel, is about 19-23 percent (Knutson, et al.). Diesel powered irrigation pumping plants fall in approximately the same range with OPEs ranging from 20-25 percent. Natural gas pumping plants have a somewhat lower range of efficiencies, 16-22 percent (Knutson, et al.).

Energy is also used directly for pumping water to farms by various irrigation organizations. Federal, state, and regional water projects and irrigation districts use pumping stations to move water along canals, lift water over mountains, and pressurize pipelines. Some irrigation districts also pump groundwater and supply it to farms through canals and pipelines. Direct energy use for irrigation in the West, for supplying water to farms, pumping from on-farm wells, and pressurizing application systems, is summarized in Chapter 8.

Additional energy is used indirectly to support irrigation. The organizations that supply surface water to farms use energy to operate control structures at dams and along canals, light and heat office buildings and workshops, operate vehicles, etc. A great deal of energy is also embodied in the facilities. Embodied energy is the energy used in the manufacture of construction materials, transportation of the materials to the construction site, operation of construction equipment, etc. Hagan and Roberts discuss the total energy requirements of water projects (direct, indirect, and embodied) and illustrate the total energy requirements of some water projects in California.

Energy is used indirectly to support on-farm irrigation application and energy is also embodied in application systems. Indirect energy use includes such things as transporting irrigation pipe into the field at the beginning of the season and back into storage at the end of the season with a truck or tractor-towed pipe trailer. The fuel used in the small engine that moves side-roll sprinklers between sets also is indirectly associated with irrigation. Embodied energy is used for manufacture of irrigation equipment, transportation to farms, installation of equipment, leveling of fields, and installing and removing head ditches, border ridges, and furrows.

The most comprehensive study made so far of manufacturing energy for irrigation system components was carried out at Oregon State University (Cuenca, et al.). The researchers included the following types of energy:

1. Primary energy--basic energy source (coal, refined oil products, natural gas, etc.) required to manufacture an irrigation component.
2. Secondary energy--energy required to produce primary energy (by mining of coal, refining of crude oil, etc.).
3. Secondary process energy--energy required to produce the process equipment.
4. Miscellaneous process energy--energy not specifically accounted for in general plant use.

The study used the process analysis approach in which flow charts were made for each manufacturing process from extraction of raw material to the fabrication of the finished product. Energy inputs at each step were quantified from published information and from consultation with manufacturing representatives. Table 3.3 lists the resultant manufacturing energy use coefficients for irrigation system materials.

Additional energy is used in the installation of on-farm wells. Methods of well drilling and energy

Table 3.3
Manufacturing energy use coefficient for irrigation sys[illegible]
materials

Material	Manufacturing Energy Requirement		
	$kwh \cdot lb^{-1}$	$kwh \cdot kg^{-1}$	$10^6 J \cdot kg^{-1}$
Aluminum Tubing			
Welded	29.65	65.37	235.32
Extruded	30.32	66.84	240.64
Weighted Average	30.09	66.34	238.81
Steel Products			
Bars	3.93	8.66	31.19
Wire	4.10	9.04	32.54
Structural Steel	3.93	8.66	31.19
Wheels	4.01	8.84	31.83
Pipe	3.95	8.71	31.35
Plates	3.92	8.64	31.11
Mean	3.97	8.76	31.53
Copper	17.46	38.49	138.57
Zinc			
Electrothermic Process	10.62	23.41	84.29
Vertical Retort Process	9.52	20.99	75.56
Electrolytic Process	8.81	19.42	69.92
Weighted Average	9.43	20.79	74.85
Brass Tubing	18.51	40.81	146.91
Low Density Polyethylene	16.75	36.93	132.94
High Density Polyethylene	16.04	35.36	127.30
Polyvinylchloride	14.65	32.30	116.27
Portland Cement			
Wet Process	1.24	2.73	9.84
Dry Process	1.11	2.45	8.81
Mean	1.18	2.59	9.33
Asbestos-Cement Pipe			
Wet Process	6.75	14.88	53.57
Dry Process	6.70	14.77	53.18
Mean	6.73	14.83	53.38

Source: Cuenca, et al.

requirements for well installation are discussed in Chapter 9.

USEFULNESS OF ENERGY ANALYSIS

The definition of energy analysis is not consistent. Various studies account for different forms of energy, including fossil fuel energy, solar energy, energy embodied in equipment, energy embodied in fertilizers and other production inputs, the energy content of agricultural products, animal power, and human labor. The different forms of energy are measured in common units such as calories or BTUs and added together. The main criticism of such energy analyses results from their use to show that certain crops contain less energy than is required to produce them, implying that such crops should not be grown. However, some of these "energy-losing" crop production activities are important human enterprises and produce economic gain for producers and useful products for consumers.

Energy analysis does have limited usefulness for contingency planning. For example, as long as our nation is partly dependent on imported oil and natural gas there remains the threat of sudden cutbacks in energy availability. Public policy makers need to know the total amount of primary (fossil-fuel) energy required for irrigation (including primary energy used to generate electricity and manufacture irrigation equipment) in order to allocate energy during an emergency energy shortage. Energy analysis in terms of primary energy will also help in devising ways of producing necessary food and fiber crops with the minimum amount of primary energy. Primary energy is used both to operate irrigation pumps and to manufacture and install irrigation systems. It might be wise public policy to encourage the installation of such things as pipeline distribution systems and automated surface irrigation systems now in order to reduce pumping energy requirements in the future. Small subsidies could be offered in the same way that many electric utilities are offering zero interest loans to residential customers for installation of better insulation and solar water heating systems.

Individual farmers usually have no need to do energy analysis on irrigation systems because their decisions are based upon costs and returns, or economic profit. However, where two or more alternative systems provide approximately equal overall net returns, the one with the lowest direct energy requirements will be affected least by future energy shortages or energy price increases. These are concerns that individual farmers may have when investing in new irrigation equipment.

COMPUTER MODEL FOR CALCULATION OF TOTAL ENERGY REQUIREMENTS OF IRRIGATION SYSTEMS

An Oregon State University research group has developed a computer model for calculation of the total energy requirements for an irrigation system. The model makes use of the manufacturing energy requirements shown in Table 3.1 and also calculates energy requirements for transportation, installation, and operation of systems. The following description of the model is taken from a report by English, et al.:

Model Description

The model evaluated the total amount of non-renewable energy consumed in the irrigation process. Energy requirements of major irrigation systems with selected crop-climatological conditions on specific acreages are calculated. The model is programmed in standard FORTRAN IV and is implemented on the Oregon State University time sharing system. The model is conversational in nature, asking questions about the irrigation system and providing answers that the modeler can use in making further decisions in the design process. The model is capable of predicting energy requirements for the following eight irrigation systems: (1) hand move, (2) center pivot, (3) drip, (4) side roll, (5) solid set, (6) surface, (7) permanent, and (8) big gun systems.

To simplify the calculation procedure, irrigation systems are divided into four basic energy consuming activities:

1) operating energy,
2) manufacturing energy,
3) transportation energy, and
4) installation energy.

The computer model is composed of 37 subprograms, groups of which simulate particular types of irrigation systems. The subprograms consist of three levels: a main program (Level I), which directs the operation of the total model; Level II subroutines, which accumulate and print final answers for a particular type of irrigation system; and Level III subroutines, which calculate energy requirements for each type of energy used. After completing the analysis of one irrigation system, the main program will initiate the analysis of another system or terminate the model as specified by the operator.

Initially the program determines which system type is to be analyzed. Program control is then transferred to the appropriate Level II subroutine (e.g., HANDMV). The Level II subroutine then calls the Level III subroutine for computing pumping energy (e.g., OPRAT1).

The total dynamic head (TDH) is found by totaling the calculated mainline friction head loss, the calculated lateral friction head loss, plus the specified input values of sprinkler operating head, pump suction lift, elevation difference from pump to field, friction loss in the suction line, and height of the riser pipe, all expressed in units of feet.

The power required to pump the water is given by the equation:

$$WHP = \frac{TDH*QPUMP}{3960}$$

where:

WHP = water horsepower (hp)
TDH = total dynamic head (ft)
QPUMP = pump discharge (gal/min)
3960 = conversion factor (ft-gal/min-hp)

To determine the brake horsepower of the motor required to drive the pump, the water horsepower must be divided by the efficiency of the pump. If an internal combustion engine is the power source, dividing the brake horsepower by the motor efficiency will yield the horsepower potential in fuel required for pumping. If an electric motor is the power source, brake horsepower must be divided by both motor efficiency and efficiency of the electric generating plant to determine the potential horsepower in fossil fuel required. The total energy required for pumping during the season is simply the product of fuel horsepower and total operating time.

A subroutine calculates manufacturing energy (e.g., MANFT1). This subroutine first calculates the energy to manufacture the mainline network. The weight per foot of tubing, the weight of each coupler, and the length of each individual pipe section comprising that segment of the mainline are used to compute the weight of each mainline segment. Multiplying this weight by the manufacturing energy per pound for the appropriate material type yields the energy to manufacture that segment. The process is repeated for each segment in the network to determine the total energy of manufacture for the mainlines. A procedure similar to that used on the mainlines is then conducted for the energy to manufacture laterals. The energy for manufacturing sprinklers is then calculated, using the assumption that all sprinklers weigh 1.1 pounds (the weight of a Rain Bird Model 30) and are entirely made of brass. The energy to manufacture the pumping plant is calculated by assuming the plant horsepower rating is the next standard size equal to or larger than the brake horsepower requirement for the pump. This unit size is chosen from a table of available motor sizes and is multiplied by a manufacturing energy per unit horsepower figure to yield the energy to manufacture the pumping plant.

Pipes are assumed to have 20 years useful life. Pumps are assumed to have 15 years useful life. Sprinklers are assumed to have 10 years useful life.

The next operation is the calculation of transportation energy (e.g., TRNSP1). Transportation energy includes energy used to move pipe to the field at the beginning of the season, to move pipe within the field during the season and to pick up pipe at the end of the season. Transportation energy is divided into two parts, manufacturing energy for the pipe trailer (if any) and fuel consumed by whatever system movement takes place. The trailer is assumed to be made of steel and to weigh a specified amount. Manufacturing energy is calculated by a process similar to those described above. Manufacturing energy is prorated over the operating life (assumed to be 20 years). Transport energy per season is presented as a single figure (by summing manufacturing energy and fuel). Manufacturing energy for the tractor used to move pipe is not included, as the prorated amount expended in moving irrigation pipes is assumed to be negligible.

System installation energy is the next major component of the program (e.g., INSTL1). For a hand move system, installation energy is assumed to be negligible unless some of the pipelines are buried. The operator has the option of specifying buried pipes. If buried, pipes are assumed to be in a trench requiring approximately one quarter of a gallon of diesel fuel per cubic yard to excavate and back fill. Pipes are assumed to have two feet of cover over them, and to require a width of four inches greater than their nominal width. The product of the total volume of excavated trench and energy per unit volume yields the total installation energy.

The total energy for seasonal operation is calculated by summing the following:

1. total seasonal pumping energy;
2. total seasonal transport energy;
3. energy to manufacture mainlines and laterals, and installation energy, all prorated over their expected life;
4. energy to manufacture pumping plant, prorated over its expected life;
5. energy to manufacture sprinklers, prorated over their expected life.

Dividing total seasonal energy by number of irrigated acres yields seasonal energy per acre. Seasonal energy per acre is divided by total seasonal application to yield seasonal energy per acre-inch.

Energy is used to transport a new irrigation system to the farm where it is to be used is then calculated using the FREIT1 subroutine. All components of the system are assumed to originate at the same point. The number of miles the system is transported by each of six types of freight systems is calculated. The types of transportation which can be specified by the operator are truck, rail, air, inland waterways, and coast and Great Lakes ship or barge transport. A combination of these types of transport is permitted by the subroutine. Energy use coefficients for freight transport are summarized below.

Truck	0.0003430 kwh/lb-mile
Rail	0.0001010 kwh/lb-mile
Inland Water	0.0000398 kwh/lb-mile
Coast/Great Lakes Ship	0.0000331 kwh/lb-mile
Coast/Great Lakes Barge	0.0000412 kwh/lb-mile

Air 0.0039600 kwh/lb-mile

All irrigation systems are modeled in a similar fashion, with some alterations to allow for basic differences between system types. For example, when a center pivot system is being modeled, the mainline is assumed to be of constant size and to run to the center of a square field. The lateral is 7 inches in diameter. The lateral and towers are assumed to be made entirely of steel. The lateral for a 160-acre field is 1,280 feet long with ten support towers each powered by a one horsepower electric motor. Tower motors are assumed to operate at three quarters of their rated capacity, and their power consumption is calculated accordingly. Sprinklers on the lateral are spaced at non-constant intervals to allow for uniform application.

When a solid set irrigation system is being simulated it must be assumed that there are enough laterals to cover the entire field, but only a portion of them operate at any one time. The segment of the mainline where laterals are in operation is treated as a manifold flow situation. Transport energy includes only that required to lay out and pick up the pipe network at the beginning and end of each irrigation season.

A side roll sprinkler system is modeled in a manner very similar to that for a hand move system. One notable exception is that only 4 and 5 inch diameter laterals are considered. The lateral walls are of heavier gauge material than standard (hand-move) laterals, and each section has a wheel as an integral part. Movement of laterals in the field is different in that a pair of laterals, one on either side of the mainline, moves as a single unit. Each of these pairs of laterals is propelled by a moving device powered by a four horsepower engine. It is assumed that the moving unit requires 5,000 kilowatt-hours of energy to manufacture and consumes one-half gallon of diesel fuel per hour of operation. It is further assumed that 15 minutes of operation per pair of laterals per move are required for transport.

A drip irrigation system is simulated in much the same manner as the solid set system. The major differences are that all laterals operate at once, and that the system is a permanent installation with buried pipelines and no required transportation energy. The operator may choose either a micro-tube type emitter or an emitter with a spiral restricting path.

For modeling a permanent type sprinkler system, a "flag" is set which eliminates the transportation subroutine since no transport energy is required. Installation energy is calculated for both buried lateral and mainline pipes. With these exceptions, the subroutines function exactly the same as when modeling a solid set sprinkler system.

For the simulation of a surface irrigation system, it is necessary to first calculate the energy for field leveling where required yardage per acre and average length of haul for leveling equipment are inputs. After calculating leveling energy, the energy to make the distribution network in the field is estimated. Two types of networks are considered, furrows and corrugations. The estimates of the energy required per acre to form furrows and corrugations were provided by Oregon farm operators. In estimating the energy required to make the field head ditch, three types of structures are considered. The available options are unlined earthen ditch, a concrete lined ditch, or a gated aluminum pipe. The earthen ditch is assumed to require a minimal amount of energy, rated at one hundredth of a kilowatt-hour per lineal foot of ditch. The concrete lined ditch is assumed to have a trapezoidal cross section with a lining 2 inches thick. The gated pipe is assumed to require approximately the same energy as aluminum mainlines of equal size, as defined in the hand move sprinkler system model. When an open head ditch is considered, devices for releasing water onto the field from the ditch can be siphon tubes or either earthen or concrete turnouts. For purposes of calculating manufacturing energy, the siphon tubes are assumed to be aluminum, 4 feet long and 1 inch in diameter, requiring about 10 kilowatt-hours per tube to manufacture. Earthen turnouts are assumed to require a negligible amount of energy, since human energy (shovelling) is the major input. Concrete turnout devices, such as gated spiles, are assumed to require 126 kilowatt-hours per structure to manufacture. It is assumed that one siphon tube is used for each furrow or corrugation, but each turnout or spile (a tube through an earthen bank) is assumed to supply water for three furrows or corrugations. The operator also has the option of using a water source that cannot be applied to the field by gravity flow. In this case the pumping energy to apply the necessary amount of water at any specified static lift is calculated. When a gated pipe is used, the friction loss in the pipe is included.

Big gun systems are assumed to involve water driven units with standard 660 foot hoses. The program computes the number of such units needed to meet the water demand of the crop. Transportation energy is embodied in pumping energy as increased head loss required to drive the unit.

Table 3.4 shows a sample of input data required by the computer program for calculation of the energy consumption of a center-pivot sprinkler system. Table 3.5 shows the

Table 3.4
Sample input to program for analysis of energy consumption by a center-pivot sprinkler system

```
ENTER POWER UNIT TYPE (0=ELECTRIC, 1=COMBUSTION) ? 0
ENTER TYPE OF LATERAL MATERIAL (1=STEEL, 2=ALUMINUM, 3=PVC,
     4=TRANSITE) ? 1
ENTER TYPE OF MAINLINE MATERIAL? 1
ENTER LATERAL LINE DIAMETER (IN)? 5
ENTER MAINLINE DIAMETER (IN) ? 5
ENTER SPRINKLER PRESSURE (PSI) ? 60
ENTER HAZEN-WILLIAMS COEFFICIENT:  MAINLINE ? 100
ENTER HAZEN-WILLIAMS COEFFICIENT:  LATERAL ? 100
ENTER STATIC LIFT (FT) ? 0
ENTER ELEVATION DIFFERENCE PUMP-TO-FIELD (FT) ? 12
ENTER SUCTION LINE FRICTION LOSS (FT) ? 0
ENTER MISCELLANEOUS FRICTION LOSS (FT) ? 10
ENTER SEASONAL APPLICATION (IN) ? 30
ENTER NET IRRIGATION REQUIREMENT (IN/IRRIGATION) ? 0.66
ENTER TOTAL LENGTH OF LATERAL LINE (FT) ? 800
ENTER LENGTH OF MAINLINE PIPE SECTION (FT) ? 40
ENTER HEIGHT OF RISER PIPE (FT) ? 0
ENTER PUMP EFFICIENCY ? .75
ENTER GENERATING PLANT EFFICIENCY ? .30
ENTER IRRIGATION EFFICIENCY ? .80
ENTER MOTOR EFFICIENCY ? .88
ENTER NUMBER OF MAINLINE SECTIONS ? 40
ENTER ROTATION TIME (HOURS) ? 36
ENTER FIELD AREA (ACRES) ? 62
ENTER NUMBER OF LATERAL SUPPORT TOWERS ? 6
ENTER NUMBER OF SPRINKLERS PER LATERAL ? 25
IS MAINLINE BURIED ? Yes
```

output of the program for the same system. This program can be obtained from its authors (English, et al.) at Oregon State University.

FUTURE IRRIGATION TECHNOLOGIES

Introduction

Considerable work has been done to improve irrigation system efficiency for water and energy savings. The technologies applied have been broad in scope. They include applications of microprocessors for control of water flow, improved estimates of crop water requirements, soil moisture monitoring methods which can be monitored remotely by a microprocessor based system, and application of newly developed irrigation equipment. In this section a few technologies will be highlighted as examples.

Table 3.5
Sample output of program for analysis of energy consumption by a center-pivot sprinkler system

```
PUMP DELIVERY CAPACITY = 643.64 GPM
LATERAL DIAMETER = 5 IN
TDH = 405.22
TOTAL OPERATING TIME 1651.00 HOURS/SEASON
NUMBER OF IRRIGATIONS PER SEASON = 46 CYCLES
ENERGY TO MANUFACTURE MAINLINES = 61804.80 KWH
ENERGY TO MANUFACTURE ROTATION LATERAL = 314463.00 KWH
DESIGN POWER UNIT CAPACITY = 100.00 HP
ENERGY TO MANUFACTURE PUMPING PLANT = 116300.00 KWH
ENERGY TO MANUFACTURE SPRINKLERS = 543.68 KWH
TRANSPORT ENERGY = 4322.08 KWH PER SEASON
IS MAINLINE BURIED  Yes
INSTALLATION ENERGY = 866.13 KWH
TOTAL PUMPING ENERGY PER SEASON = 410764.35 KWH
TOTAL SEASONAL OPERATING ENERGY = 415086.43 KWH
TOTAL SEASONAL MANUFACTURING ENERGY = 26664.40 KWH
TOTAL SEASONAL ENERGY = 441750.82 KWH
SEASONAL ENERGY PER ACRE = 7126.01 KWH
ENERGY PER ACRE-INCH = 237.50 KWH/ACRE-INCH
```

When a new technology is both needed and cost effective farmers have been quick to adapt it. Systems which were only in the conceptual stage a few years ago are now being extensively applied in the field. Center-pivot systems which are clearly visible in satellite photographs are an obvious example of how a technology works. Application of low pressure technology which reduces sprinkler operating pressure from 60 to 80 psi to 15 to 25 psi is a response to the rising cost of energy. Another modification of center-pivot systems has been the use of tubes which hang from the pivot arm and transmit water to a spray nozzle set at a height just above the crop canopy. When such nozzles are used in conjunction with row crops having furrow diking, the system is referred to as a low energy precision application or LEPA system.

Research on remote sensing has led to applications of infrared temperature measurement methods on crop canopies as an attempt at irrigation scheduling. This technology is not yet fully developed or ready for widespread adoption, however. Technologies will only be applied if a number of factors all come into line at the same time. This includes operator understanding of the system and ability to maintain the system. The technology must be cost effective and generally provide an improvement in the efficiency of water or energy use or increase crop output.

Drip/Micro Irrigation Systems

Drip irrigation systems have been applied mostly to widely spaced crops such as orchards and vineyards. Drip continues to be used in orchards and has been found effective to promote crop production and quality in the early stages of orchard development. These drip systems use a single emitter or group of emitters to serve a single plant.

Some drip irrigation is now being applied to row crops. This requires a new type of drip application system normally involving long lengths of tubing running along the crop row with water application being more dense than with single or multiple emitters. Such water distribution is accomplished by frequently spaced inline emitters or by bi-wall tubing. Bi-wall tubing contains a perforated outer tube which is connected to an inner tube which transmits the water along the line by another set of perforations. Such tubing is normally produced sufficiently inexpensive that it may be discarded after a relatively few years, perhaps three to five, and replaced by a new tubing. Any drip system is subject to clogging by silt and bacterial slime which are difficult to remove. Such material must be prevented or removed by proper filtration systems and even by application of chlorine treatments.

Micro-irrigation refers to systems which apply irrigation water close to the crop canopy or soil surface by way of small diameter tubes. Unlike common drip systems, these tubes may be continuously moved through a field by use of a center-pivot or lateral-move system which distributes the water to the individual tubes. Micro-irrigation is a potential technology which attempts to combine the labor effectiveness and operation flexibility of a center-pivot or lateral-move system with the high application efficiency and low evaporation loss of a drip irrigation system. Research is continuing on micro-irrigation systems under field conditions.

Surge Irrigation

For many years, researchers have tried to increase efficiency and further automate surface irrigation methods. One such method is surge irrigation which is applicable in surface irrigated row crops. The concept of this method is to apply water at a relatively high rate of flow down a furrow for a relatively short period of time. The water flow is then switched off one part of the field by an automatic valve and distributed to another part of the field. The furrows are alternately wetted and allowed to dry by the pulses of water. This process causes increased application efficiency due to the physical process of infiltration to the soil surface, particularly during the first few irrigations of the growing season.

Surge flow irrigation may fit well into many furrow irrigation systems. The cost of the automatic surge valve is relatively low and it is dependable in field trials. More manufacturers are beginning to produce surge irrigation equipment, particularly in those areas where the land is predominantly surface irrigated.

Irrigation Scheduling

There are numerous ways to schedule irrigations and estimate the required depth of water application.

Sophisticated or scientifically based irrigation scheduling methods which require either mathematical estimation of crop water use or a measurement of soil moisture content have been available for decades. However the general application of irrigation scheduling methods has been sporadic but is increasing in response to needs for raising the efficiency of using water and energy. Modern technology has enhanced the potential application of irrigation scheduling methods by developing microprocessors to monitor sensors for measuring crop water needs and computers to aid in the management decisions of irrigation scheduling.

Microprocessors are used in automated weather stations which have sensors for temperature, relative humidity, wind speed, and solar radiation. Such information can be conveniently collected and stored by a microprocessor based system and then transmitted to a satellite receiving station where the information can be beamed to a central computer. The meteorological data taken by the sensors are used in formulas to estimate evapotranspiration or crop water use for a wide range of crops. Such systems have been applied to practically all the western states for use in irrigation scheduling.

Additional equipment is available for use in soil moisture monitoring which can also be a basis for irrigation scheduling. Soil moisture monitoring may be done by a soil-moisture tension measuring device such as gypsum blocks. The soil-moisture tension can be used directly to efficiently schedule irrigations for maximum crop production or water use efficiency.

The neutron probe is another relatively new technology that has been applied to irrigation scheduling. The neutron probe can be used to monitor soil-moisture content through the root profile. The criterion of limiting level of soil-moisture content can then be used to schedule irrigations. The neutron probe can quantify the amount of water removed over the soil profile and therefore help to determine the required depth of irrigation. Such devices are being increasingly used by large scale irrigation scheduling services which have responsibility for monitoring large acreages of diverse crops over many farms.

REFERENCES

Addink, J.W., J. Keller, C.H. Pair, R.E. Sneed, and J.W. Wolfe. "Design and Operation of Sprinkler Systems." In Design and Operation of Farm Irrigation Systems, M.E. Jensen, ed., American Society of Agricultural Engineers, pp. 628-660, 1980.

American Association for Vocational Instruction Materials. "Planning for an Irrigation System." 1971.

Cuenca, R.H., M.J. English, and G. Daskalakis. "Energy for Manufacture of Irrigation Systems." Appendix B in Analyses of Energy Used by Irrigation Systems, by B.J. English, R.H. Cuenca, Kuei-Lin Chen, R.B. Wensink, and J.W. Wolfe, Agricultural Engineering Department, Oregon State University, June 1982.

Daubert, J. and H. Azer. "Laser Levelling and Farm Profits." University of Arizona Tech. Bull, B-224, 1982.

English, B.J., R.H. Cuenca, Kuei-Lin Chen, R.B. Wensink, and J.W. Wolfe. "Analyses of Energy Used by Irrigation Systems." Agricultural Engineering Department, Oregon State University, June 1982. Appendix A: Description of the Model.

Erie, Leonard J. and Allen R. Dedrick. "Level-Basin Irrigation: A Method for Conserving Water and Labor." U.S. Department of Agriculture, Farmers Bulletin No. 2261, April 1979.

Fereres, E., J.L. Meyer, F.K. Aljibury, H. Schulbach, A.W. Marsh, A.D. Reed. "Irrigation Costs." University of California, Division of Agricultural Sciences Leaflet 2875, August 1978.

Hagan, Robert M. and Edwin B. Roberts. "Energy Impact Analysis in Water Project Planning." Journal of the Water Resources Planning and Management Division, ASCE, 106(WR1): 289-302, 1980.

Hart, W.E., H.G. Collins, G. Woodward, and A.S. Humpherys. "Design and Operation of Gravity or Surface Systems." In Design and Operation of Farm Irrigation Systems, M.E. Jensen, ed., American Society of Agricultural Engineers, pp. 501-580, 1980.

Knutson, Gerald D., Robert G. Curley, and Blaine R. Hanson. "Cutting Energy Costs for Pumping Irrigation Water." University of California, Division of Agricultural Sciences Leaflet 21188, 1981.

Marshall, Harold E. and Rosalie T. Ruegg. "Simplified Energy Design Economics." U.S. Department of Commerce, National Bureau of Standards, NBS Special Publication 544, January 1980.

Merriam, J.L. "Surface Irrigation Methods." In Agricultural Water Conservation Conference Proceedings, University of California, Davis, pp. 62-81, June 23-24, 1976.

Merriam, J.L., M.N. Shearer, and C.M. Burt. "Evaluating Irrigation Systems and Practices." In Design and Operation of Farm Irrigation Systems, B.E. Jensen, ed., American Society of Agricultural Engineers, 1980, pp. 721-760.

Pair, C.H., Editor-in-Chief. Sprinkler Irrigation. Fourth Edition. Sprinkler Irrigation Association, 1975.

Shearer, Marvin N. "Comparative Efficiency of Irrigation Systems." In Efficiency in Irrigation. The Irrigation Association 1978 Annual Technical Conference Proceedings, pp. 183-188, February 26-28, 1978.

U.S. Soil Conservation Service. Border Irrigation. National Engineering Handbook, Section 15, Chapter 4, 1974.

4
Efficiency and Optimization in Irrigation Analysis

James C. Wade

INTRODUCTION

Conservation, Efficiency and Optimization

Conservation is the wise use of society's natural resources for present and future generations. Through individual decisions, negotiations, and legislation society makes resource choices and sets conservation policies. In this context, conservation decisions reflect society's attitudes toward and strategies for using natural resources in production, recreation, and culture. Conservation motivations lead to an array of alternative resource management strategies. Among the strategies often discussed are those that improve "efficiency" of water and energy use in irrigation. But what is meant by efficiency in irrigation and how does one use efficiency in decision making?

The societal goal, to be efficient, is to produce more output with less resources. The two basic representations of efficiency are physical (also called technical) and economic. Physical efficiency measures the output of production relative to the input in physical terms. Economic efficiency, a somewhat more complex measure, evaluates the economic output or "benefits" of production relative to economic costs. Optimization is simply seeking the alternative with the "best" value of the efficiency measure.

In a world of complete knowledge, the prices and supplies of scarce resources are sufficient to determine the economically efficient use of resources over time.

However, the existing world of incomplete information, imperfect competition, property rights, and complex resource externalities makes resource planning, policy and programs essential for reflecting the goals of society. Thus, the composition and extent of resource scarcity and the strategies for resource use and conservation are important topics of politics, conversation, and study as society seeks to provide for current and future needs. Social attitudes, as well as actual resource strategies, are reflected in the policies and programs initiated by local, state, and federal governments.

Farmers using irrigation apply their knowledge of production technology and the prices of resources to develop a resource use plan appropriate to their individual objectives. At the firm level, resource conservation is carried out through decisions about personal utility and/or firm profits. Water and energy will be conserved to the extent that it is consistent with the goals of utility or profit maximization.

Choosing Irrigation Systems

Irrigation management decisions are made in an environment of technical feasibility considering the quantity and quality of the resources available on the farm. These resources may include soil, climate, water, topography, labor, capital, and management skills. The relationship between various technical alternatives, their resource use, and the production constraints placed on the farm leads each farmer to choose the most appropriate irrigation technology.

Alternate irrigation methods may range from the most primitive gravity flow irrigation systems to the latest pressurized application systems. Other choices a farmer might consider are conventional or laser leveling of fields, the size and shape of fields, the length of water run, the timing and scheduling of water application, and the crops to be grown with a particular irrigation technology. Each irrigation system utilizes different combinations of labor, capital, water, energy, and other resources. The choice of an irrigation system might also be restricted by the land topology, soil texture, climate, crops, and water availability.

Recent irrigation system studies have emphasized the concepts of irrigation scheduling, peak load irrigating, and deficit irrigation (Bernardo and Whittlesey; Dudek and Horner; Harris and Mapp; and Taylor, et al.). Irrigation scheduling is a management technique designed to apply water to a crop in optimum quantities and at proper times to produce crops in the most efficient manner. Such procedures attempt to reduce water losses due to soil evaporation and percolation by supplying water only when and where it is needed. The criteria for scheduling irrigations usually are designed to produce a near maximum crop yield unless water scarcity is a problem. Peak load energy management involves using energy in "off-peak" hours when the total demand for energy is lowest. For accepting an interruptible supply of energy, irrigators receive a reduced rate for purchased energy. Several variations of energy cost management are followed, but usually determined by the method of pricing energy by local utilities.

Deficit irrigation schemes may utilize any technique of water application (Bernardo and Whittlesey; Taylor, et al.). However, irrigation water is deliberately held to levels less than needed by the plant to obtain maximum yields. Limited water application schemes produce less output with less water and, therefore, at lower total revenue and cost. Net economic returns may be increased if water or energy costs are very high. However, this approach is more likely followed in response to physical restrictions on supplies of water and/or energy.

Efficiency and optimization are important concepts in determining resource use. The choice of efficient irrigation systems is a primary concern of farmers. Although policy makers and agricultural researchers are concerned with efficient resource use, the farm level decision maker ultimately chooses the design and use of on-farm irrigation systems. Policy makers and researchers must provide incentives and information to aid the farmer in making these decisions.

EFFICIENCY AND OPTIMIZATION

Statements concerning the relative efficiency of alternative production processes are often confusing and misleading. The difficulty in measuring and describing efficiency lies in the alternative ways the concept can be used. Several alternative definitions of efficiency are presented in this section.

Defining Efficiency

Efficiency of resource use is measured, in its simplest form, as the ratio of outputs to inputs for a production process. Each measurement of efficiency must be placed into context (Spedding, et al.). The context of an efficiency measure is defined by (1) limiting the system for which efficiency is measured, (2) specifying the outputs and inputs of concern, and (3) describing the time period over which the system is analyzed. The desirability of one level of efficiency over another is subjective and primarily left to the decision maker who must examine or use the measure.

In this chapter, efficiency measured in terms of output to input ratios is called "output efficiency." The most common form of this measure forms a ratio between a single input such as irrigation water or energy and the marketable output of the system. Other inputs used in production may be assumed unchanged or nonlimiting when measuring the output efficiency of a single input. Output efficiency is measured in units such as bushels per acre, tons per acre foot of water, or miles per gallon. Economists call output efficiency the average physical product of the input.

If a choice is to be made among systems having the same levels of output and if one of the systems is a standard by which all other systems can be judged, the "input efficiency" of the alternative system can be measured as the ratio of its input requirements to the input requirements of the standard system. Although it is assumed that the levels of all other inputs are the same for both systems, such is rarely the case.

On the surface, the implications of output and input efficiency are very simple: a producer or society wants to get as much output as possible from the inputs available. A closer examination of a particular situation shows that a gain in the perceived efficiency of an input may be a loss in "efficiency" of another input or in another part of the system. That is, since production of an output always requires more than one input, what may be efficient for one input may be inefficient for another, measured in physical terms. Hence, the question arises, How does one adequately evaluate efficiency? It seems that more complete efficiency measures are needed. Additionally, some "value" or "purpose" must be associated with the measure.

In irrigated farming, land, labor, equipment, water, electricity, seed, fertilizer, and many other inputs are used to produce crops. Research develops new plants, methods, and equipment to combine inputs in different ways. Changing from one production method to another may or may not affect the use of an input. For example, in examining alternative irrigation methods, the use of seed, fertilizer, labor, and land may not change drastically from one method to another. However, inputs like water and energy associated with irrigation may change significantly.

If more than one technology is considered for producing the same product, inputs can potentially be utilized in many combinations to produce any level of output. Any one of these input combinations can be economically efficient. However, some technologies may be more physically efficient in the use of some resources, like water or energy, than others. Adoption of a technology to raise the efficiency of using one input may create an ambiguity, since decreasing the use of one input may require an increase in the use of other inputs.

Efficiency measures examine the physical aspects of alternative sets of inputs but leave the question of identifying which set is superior to the decision maker, a process of optimization. This can be done by simply choosing the alternative with the highest overall economic efficiency, minimizing the total cost of all inputs required to produce a given output. The value of nearly all inputs and outputs is determined in a market setting. Economics provides a set of allocative measures for choosing input and output levels for a technological process.

Economic measures (costs, revenue, profits, etc.), like the physical measures of efficiency, are evaluations of alternative input/output situations and, as such, do not imply one level of a measure is superior to another. To establish an ordering of measures, choice criteria are used to analyze the measures for decision making. The criteria usually lead to some form of optimization.

Defining Optimization

In regard to the use of efficiency measures, Ladd has noted that

> One measure of efficiency cannot be proven superior to another measure without some (perhaps implicit) prior

> assumption about criteria for desirable efficiency measures. . .(Ladd, p. 1).

From this statement, one notes that efficiency can be evaluated only in the context of a preconceived idea of what is "better" or "best." Hence, the concepts of efficiency and resource conservation can have different meanings to different individuals at different times.

Farmers, researchers and policy makers each have concepts of optimality. For farmers, "better" often relates to the economic and social well being of the farm firm or family. These goals are generally centered in economic efficiency and profits. Policy makers' criteria, on the other hand, tend to change as elections decide the path of social change. Such goals concern broader strategic and long term resource management objectives. For example, the goal of producing commodities with a minimum energy input may not be transferable to individuals who are motivated only by economic concerns. In a free market economy most individual decision makers watch for signals, usually prices, to guide their use of resources.

Technology developments often respond to the goals and pressures imposed by social policy and their perceptions of farmer goals. To improve an irrigation method by reducing the energy or water used, a researcher might construct motors, pumps, applicators or even develop new irrigation technology. The efficiency measure, in this case, is reducing an input use relative to a previously established standard. The standard may be imposed by others or simply provided by historical observation. The criteria are often vague and conflicting; as in the case that less energy use is always superior to more energy use. Such physical efficiency measures may be in conflict with the more frequently applied criteria of economic efficiency in individual decisions.

Optimization and efficiency are related concepts. Optimization seeks to attain the "best" situation in terms of stated criteria. The constraints imposed on the choice can reflect resource limitations or normative restrictions imposed by an individual. An optimal solution is the "most" efficient in terms of the criteria chosen. Maximum efficiency results from the decisions implied by an optimization process. Both Spedding and Ladd indicate that efficiency measures cannot be compared without some preconceived set of criteria. The choice of the criteria remains, to a large extent, a value judgment.

QUANTIFICATION OF EFFICIENCY IN IRRIGATION

Water Use Efficiency

Conservation of energy and water in irrigated agriculture is largely based on increasing efficiency of using these inputs. Farmers, engineers, economists, and resource planners are all concerned with efficiency, as each perceives it, in the allocation of resources in irrigation. Modern irrigation technologies present resource users and analysts with several unique opportunities for modifying input use and input use efficiency.

Water use efficiency in irrigation is usually measured as the ratio of the water used by a plant (transpiration) plus the water evaporated from the soil surface (evaporation) to the amount of water used for irrigation. This ratio measures the degree of technical efficiency of water use. Water use efficiency may be measured for a field, farm, or an entire irrigation district.

Sources of water use inefficiency include seepage and evaporation from main canals, ditches and other on-farm delivery systems plus deep percolation and runoff from the fields. Each of these sources of loss are also a potential source of water savings and improved water use efficiency as they are reduced or eliminated.

Water use efficiency in broad context has several components. Israelsen, et al; Michael, Bos and Nugteren; and Halderman each list definitions for the components of water use efficiency which are widely used in the irrigation literature and by irrigation specialists. Other analyses are provided by Hagan, Haise and Edminster; U.S. Department of Agriculture; and Peri, Skogerboe and Karmal. In an idealized irrigation system, water would be delivered from the source to the crop without loss. However, the technical nature of irrigation is such that water must be stored, transported and applied in ways that are less than physically ideal. Resource efficiency is sacrificed for economic gain. Various environmental parameters such as soil, topography and weather in addition to physical characteristics of the irrigation system, may affect the efficiency of water use.

Table 4.1 summarizes some existing definitions of water use efficiency. Each measure compares water use to established criteria or goals, thus, measuring the degree of technical efficiency of some aspect of water use. Water

Table 4.1
Irrigation efficiency formulas, percent

Israelsen	Michael	Bos and Nugteren	Halderman
Water Conveyance Efficiency:	Water Conveyance Efficiency:	Water Conveyance Efficiency:	Conveyance Efficiency:
$E_c = \frac{W_f}{W_r}$	$E_c = \frac{W_f}{W_d} \times 100$	$e_c = \frac{V_f}{V_t}$	$E_c = \frac{\text{water delivered to field}}{\text{water diverted from reservoir or pumped from the well}}$
W_f = water delivered to the farm W_r = Water diverted from river or reservoir	W_f = water delivered to the irrigated plot (at the field supply channel) W_d = water diverted from the source	V_f = volume of water delivered to all farm or group inlets V_t = total quantity of water supplied to the area	
Water Application Efficiency:	Water Application Efficiency:	Field Application Efficiency:	Application Efficiency:
$E_a = 100 \frac{W_s}{W_f}$	$E_a = \frac{W_s}{W_f} \times 100$	$e_a = \frac{V_n}{V_a}$	$E_a = \frac{\text{water stored as moisture in root zone}}{\text{water delivered to field including sources of losses}}$
W_s = water stored in the soil root zone during irrigation W_f = water delivered to the farm	W_s = water stored in the root zone of the plants W_f = water delivered to the field (at the field supply channel)	V_n = rainfall deficit (difference between consumptive use and effective rainfall over the cropped area)	E_a = water delivered − surface runoff + deep seepage + evaporation)

V_a = field application to the cropped area

Considering common sources of irrigation water loss

R_f = surface runoff

D_f = deep percolation below the farm root zone soil

$W_f = W_s + r_f + D_f$

so,

$$E_a = 100 \frac{W_f - (R_f + D_f)}{W_f}$$

<u>Water Storage Efficiency:</u>

$$E_s = 100 \frac{W_s}{W_n}$$

W_s = water stored in the root zone during irrigation

W_n = water needed in the root zone prior to the irrigation

<u>Water Storage Efficiency:</u>

$$E_s = \frac{W_s}{W_n} \times 100$$

W_s = water stored in the root zone during irrigation

W_n = water needed in the root zone prior to irrigation

<u>Consumptive-Use Efficiency</u>

$$E_{cu} = 100 \frac{W_{cu}}{W_d}$$

(Continued)

Table 4.1 (Cont.)

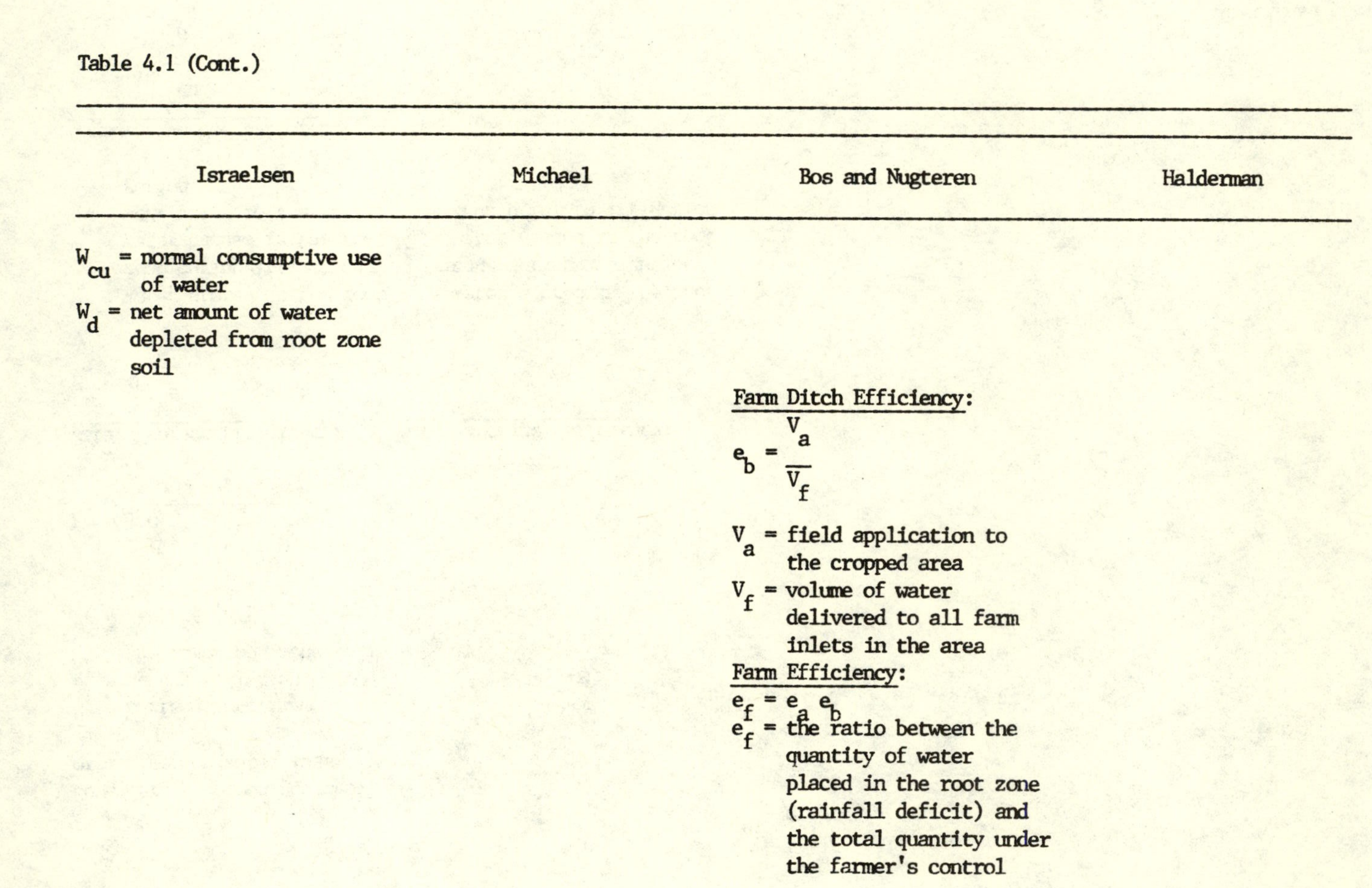

Israelsen	Michael	Bos and Nugteren	Haldeman
W_{cu} = normal consumptive use of water W_d = net amount of water depleted from root zone soil			
		<u>Farm Ditch Efficiency:</u> $e_b = \frac{V_a}{V_f}$ V_a = field application to the cropped area V_f = volume of water delivered to all farm inlets in the area	
		<u>Farm Efficiency:</u> $e_f = e_a e_b$ e_f = the ratio between the quantity of water placed in the root zone (rainfall deficit) and the total quantity under the farmer's control	

Water Distribution Efficiency: $E_d = 100\ [1-(y/d)]$ y = average numerical deviation in depth of water stored from average depth stored during the irrigation d = average depth of water stored during the irrigation	**Water Distribution Efficiency:** $E_d = (1-\bar{y}/\bar{d}) \times 100$ $\bar{d}$ = average depth of water stored along the run during the irrigation $\bar{y}$ = average numerical deviation from $\bar{d}$	**Distribution Efficiency:** $e_d = \frac{V_a}{V_t} = e_b\ e_c$ e_d = the ratio between the quantity of water applied to the fields and the total quantity supplied to the irrigated area	
Water Use Efficiency: $E_u = 100 \frac{W_u}{W_d}$ W_u = water beneficially used W_d = water delivered	**Water Use Efficiency:** This efficiency is expressed in the following ways: i) Crop water use efficiency: water use efficiency = $\frac{Y}{E_T}$ Y = crop yield E_T = water depleted by the crop in the process of evapotranspiration		**Water Use Efficiency:** $E_u = \frac{\text{water beneficially used}}{\text{water delivered}}$ beneficial water use includes: -leaching salts -subbing moisture for germination -temperature control -fertilizer management and pest control -facilitate harvesting

(Continued)

Table 4.1 (Cont.)

Israelsen	Michael	Bos and Nugteren	Halderman
	ii) Field water use efficiency Field water use efficiency $= \frac{Y}{W_R}$ Y = crop yield W_R = amount of water used in the field		For lack of better information, the consumptive use requirement for maximum yield, divided by water delivered, is often used to indicate efficiency. However, this ratio does not credit the system for beneficial uses other than consumptive use. Also, it provides little information about the efficiency of each irrigation except that yields are indicative of the water management.
	<u>Project Efficiency</u>: Indicates the effective use of irrigation water source in crop production—it is	<u>Overall (or project) Efficiency</u>: $e_p = \frac{V_n}{V_t} = e_a e_b e_c = e_a e_d$	

the percentage of irrigation water that is stored in the soil and is available for consumptive use by crops—when delivered water is measured at the point of diversion from the canal or the main source of	$= e_f e_c$ = the ratio between the quantity of water placed in the root zone (rain deficit) and the total quantity supplied to the irrigated area The overall (or project) efficiency represents the efficiency of the entire operation between diversion or source of flow and the root zone. By taking the complementary value, one can obtain the total percentage of losses.
<u>Operational Efficiency</u>: Ratio of the actual project efficiency compared to the operational efficiency of an ideally designed and managed system using the same irrigation method and facilities.	

(Continued)

Table 4.1 (Cont.)

Israelsen	Michael	Bos and Nugteren	Halderman
	Economic (irrigation) Efficiency: Ratio of the total production (net or gross profit) attained with the operating irrigation system, compared to total production expected under ideal conditions. This parameter is a measure of the overall efficiency, because it relates the final output to input.		
			Irrigation Adequacy: Adequacy Index= $\frac{\text{Soil moisture stored during an irrigation}}{\text{soil moisture deficiency}}$ measures the degree to which the soil moisture reservoir is replenished.

<u>Irrigation Uniformity:</u>

$$C_u = 1 - \frac{(\text{sum of deviations}) / \text{number of points}}{\text{average application}}$$

Irrigation uniformity is related to efficiency. Uniformity coefficient has been used chiefly for sprinkler irrigation but the concept is also useful for surface irrigation.

Any efficiency term should represent output divided by input

$$\text{Efficiency} = \frac{\text{Output}}{\text{Input}}$$

delivery efficiency compares the water output of the delivery system to the water input to the system. The range of definitions given in Table 4.1 shows the implications of considering a single plant, a farm, an irrigation district, or an entire river basin.

All of these components are important and may indicate potentials for saving quantities of water, most are not very complex. However, field application efficiency offers the major challenge for on-farm management decisions.

In observing the measures of irrigation efficiency in Table 4.1, it is apparent that not all definitions are alike. Briefly, water conveyance efficiency measures the ratio of water delivered to a farm or field in comparison to the amount of water diverted from its source. Conveyance efficiency reflects the water lost to seepage, evaporation, or spillage from the water delivery system between the point of diversion and the point of delivery.

Water application efficiency is generally considered to be a ratio between the amount of water stored in the soil root zone of the crop and the amount of water applied to the field. Water lost through percolation beyond the root zone, field runoff during application, and evaporation during application is reflected in this efficiency measure. It is usually assumed that the amount of water stored in the root zone of plants is also the amount consumed through the evapotranspiration process of plant growth. Application efficiency is of prime importance when considering irrigation management or technological adjustment to affect water use levels.

Water storage efficiency is a measure not frequently used. It is a ratio between the amount of water stored in the root zone during irrigation and the amount of water needed in the root zone prior to irrigation. In some respects, the higher the ratio, the more likely it is that water will be wasted through field runoff or deep percolation. So it is not a measure of efficiency as directly useful in the concern about water or energy conservation as are some of the others.

Consumptive use efficiency measures the relationship between normal consumptive use of water by the crop and the net amount of water depleted from the root zone. This efficiency measure is also very specialized in its use and is not generally an indicator of resource use efficiency. Farm ditch efficiency is much like the measure of conveyance efficiency described earlier.

The remaining portions of Table 4.1 are self evident in their descriptions of irrigation efficiency. It is apparent that technical resource use efficiency can be measured in many ways. It is always important to understand the definition of an efficiency measure and the context in which it is being used.

Other Basic Principles

Numerous publications have compared costs for different irrigation methods, usually without reference to crop types, soil types, or topography. Magazine articles and advertising literature frequently cite relative costs for different irrigation methods without explaining the assumptions used in calculating the costs. These types of information have limited usefulness because irrigation costs with each method are highly dependent on crop and site conditions, topography, water sources, field size and shape, and other factors. Obviously, when the costs or efficiency measures of two or more irrigation systems are compared, crop and site conditions and water sources must be the same for each.

The systems also should be designed to provide the crop similar amounts of water and, hopefully, produce the same crop yield. However, because of the different ways in which irrigation systems apply water, it is not possible to make all systems completely comparable. For example, border, furrow, and drip irrigation systems apply water directly to the soil surface. Sprinkler systems discharge water into the air and part or all of the crop foliage is wet by the spray. The changed microclimate of an area irrigated by sprinklers is likely to have some effect on yield even if the net application of water to the root zone of the crop is the same as it would be with surface irrigation. Effects on yield can be taken into account in economic analyses but they are difficult to quantify. The comparability of different types of irrigation systems is also affected by frequency of application, adequacy of irrigation, and the use of related resources such as energy and labor.

For an irrigation system, increasing field application efficiency reduces the application of nonessential water to irrigated fields. Nonessential water, although lost to the crop, is not necessarily lost to the overall water resource system. Part of the water percolates into the groundwater reservoir and is potentially available for reuse at another time or place. Additional water in excess to the plant requirements (consumptive use) may run off into local streams and be available for reuse. These "losses" may become externalities that benefit or harm other water users.

Each technology such as flood or sprinkler systems applies irrigation water with some application loss. Application losses normally take the form of field run-off and deep percolation through the root zone. As application losses approach zero, the field application efficiency approaches one.

Israelsen points out that cost and quantity of labor, ease of handling water, crops being irrigated, and soil characteristics influence efficiency. Other conditions include topography, capital requirements, management skill, energy requirements, and weather. Personal preferences and availability of information might be added to the list. Skillful irrigation management and appropriate technological adjustments can minimize the adverse impacts of these conditions upon irrigation water use efficiencies.

The characteristics of irrigation systems can be utilized to improve efficiency by applying water more directly to the point of need, reducing application losses, and improving the timing of water application. Some of the variable site and situation factors that affect field application efficiency of various systems are listed in Table 4.2. The adjustments in irrigation equipment and/or management practices required to improve field application efficiency may include new equipment, additional labor, or improved management. Increased efficiency may require improving catchment, return flow, delivery systems, overall water system management and many other related activities.

Water use efficiency (or any other single technical efficiency criterion) is limited as a sole condition for examining irrigation systems. Maximizing water use efficiency on a single field, a farm, an entire irrigation district, or river basin considers water as the most important input by disregarding potential trade-offs with other inputs. In reality, all of these items must be considered to determine an appropriate water management system.

Energy Efficiency

Efficiency as previously defined also applies to energy use. Energy use efficiency measures the ratio of an activity output to the energy input for the activity. Assessing energy use efficiency can be both simple and complex

Table 4.2
Characteristics of alternative irrigation systems with variable site and situation factors

Site and Situation Factor	Surface Systems		Sprinkler Systems			Drip Systems	
	Redesigned Surface Systems	Level Basins	Intermittent Mechanical Move	Continuous Mechanical Move	Solid Set	Emitters & Porous Tubes	Bubblers and Spitters
Average Efficiency Rating	60-70%	80%	70-80%	80%	70-80%	80-90%	80-90%
Soil	Uniform soils with moderate to low infiltration	Uniform soils with moderate to low infiltration	All	Sandy or high infiltration rate soils	All	All	All, basin required for medium and low intake soils
Topography	Moderate slopes	Small slopes	Level to rolling	Level to rolling	Level to rolling	All	All
Crops	All	All	Generally shorter crops	All but trees and vineyards	All	High value required	High value required

(Continued)

Table 4.2 (Cont.)

Site and Situation Factor	Surface Systems		Sprinkler Systems			Drip Systems	
	Redesigned Surface Systems	Level Basins	Intermittent Mechanical Move	Continuous Mechanical Move	Solid Set	Emitters & Porous Tubes	Bubblers and Spitters
Water Supply	Large streams	Very large streams	Small streams nearly continuous	Small streams nearly continuous	Small streams	Small streams continuous and clean	Small streams continuous
Water Quality	All but very high salts	All	Salty water may harm plants	Salty water may harm plants	Salty water may harm plants	All, can potentially use high salt waters	All, can potentially use high salt waters
Labor Requirement	High, training required	Low, some training	Moderate, some training	Low, some training	Low to high, little training	Low to high, some training	Low, little training
Energy Requirement	Low	Low	Moderate to high	Moderate to high	Moderate	Low to moderate	Low

Management Skill	Moderate	Moderate	Moderate	Moderate to high	Moderate	High	High
Machinery Operations	Medium to long fields	Short fields	Medium field length, small interference	Circular fields, some interference	Some interference	May have consider- able interference	Some interference
Duration of Use	Short to long term	Long term	Short to medium term	Short to medium term	Long term	Long term, but durability unknown	Long term
Weather	All	All	Poor in windy conditions	Better in windy conditions than other sprinklers	Windy conditions reduce per- formance; good for cooling	All	All

depending upon the ways a system uses energy. For example, the commonly used measure of miles per gallon of gasoline can be used to compare the relative gasoline use of two automobiles. If the criteria of choice is the quantity of fuel to be used to traverse a given distance, one vehicle can be declared superior to another. However, the overall context of purchasing an automobile also includes its size, comfort, style, and cost. Rarely is a decision to purchase an automobile made on the single criterion of energy use. Similarly, irrigation systems are not chosen only for their water or energy use efficiency.

Energy use efficiency is affected by the ability of a pump or motor to convert consumed energy into a form that lifts water. In this context, energy input and output of engines, motors, or other energy conversion processes can be measured in some primary form such as BTU (British Thermal Units). The energy use efficiency of pumps is a primary factor in the cost of irrigation water. The typical farmer or decision maker is not concerned about the technical aspects of converting one form of energy into another but only with the cost, performance, and dependability of the alternatives for choice.

Like all measures of efficiency, energy use and energy conversion efficiencies provide for comparisons of irrigation systems or components of systems. But relative merit of alternative systems must be measured against criteria such as firm profits, growth, or security. Energy use efficiency is only one of several measures that might enter the decision process of irrigation management. Economic efficiency or farm profits are most likely to be the final determinant of water and energy input levels.

Energy use, in contrast to water use, takes on several forms in an irrigation system. For example, energy in the form of fuels consumed by the engines or motors used to drive the pumps is direct energy. The energy required to produce the pumps, pipes, and other equipment is embodied energy. Energy used in the production of other inputs such as fertilizer and pesticides is an example of indirect energy use. The energy of the sun is yet another form of energy. Energy analysis is of limited use as a decision criterion at any level since its basic assumption is that energy is the primary measure of value (Edwards).

SYSTEM OPTIMIZATION

Level of the System

Optimization involves choosing an alternative with the "best" performance according to a set of selection criteria. The level or extent of the system to be optimized is a critical aspect in defining approaches to optimization. The level of a system defines its technologies, resources, and the set of alternative criteria appropriate for decision making. From the least to the most aggregated system, optimization of water and energy resource use can be accomplished at the following levels: component of an irrigation system, irrigation system, farm, region, and nation. Several individual or multiple optimizing criteria might be applied at each level of a system. Optimization may be less difficult at either end of the continuum of aggregation; at the disaggregate end because criteria and the system are generally well defined and at the aggregate end because of simplifying assumptions. Other levels offer more explicit specification of the optimizing conditions which increase the complexity of analysis.

The level of system chosen for analysis, in part, defines the conditions of optimization. However, time and uncertainty may be explicitly defined for each choice of system in light of the intent of the analysis. Optimization at any level of aggregation may involve static, dynamic, and/or stochastic elements. Static optimization elements are those that do not change with time. If all components of a system are fixed and the analysis is carried out for a single time period, the optimization is static. In some cases, the static assumptions are too simplistic to incorporate many important aspects of choice. Obviously, systems having elements that change with time are dynamic systems. Stochastic conditions are those that are not known with certainty but can be defined using probability. Both static and dynamic components may be stochastic. If system components are assumed to be known with certainty, optimization is nonstochastic or deterministic.

Optimization Criteria

The most fundamental aspect of optimization is the choice of criteria or objectives. Most producers are interested in maximizing physical efficiency only if it is consistent with other goals such as maximizing profits or family benefits.

Any water use efficiency measure could be chosen for optimization. Alternate irrigation systems or techniques could be evaluated to maximize water application efficiency. Under such a criterion the water application efficiencies of alternate systems would be compared and the system with the highest efficiency selected. Imposing specific conditions such as soil type, topography, or budget constraints may reduce the set of alternatives. The problem becomes one of maximizing one criterion subject to constraints for other measures.

Water use efficiency is compatible with economic criteria for irrigation system optimization to the extent that (1) water is expensive and/or of limited supply, and (2) the cost or supply of other inputs associated with water (such as energy for pumping or fertilizer) is of concern. Maximizing water use efficiency in irrigation may or may not be an acceptable goal. Water use efficiency rarely stands alone as a criterion for irrigation system optimization.

The conditions which limit energy use as a goal for optimizing an irrigation system are similar to those listed for water. However, there is a major difference because energy can enter a production process in numerous ways: direct, indirect, embodied, and incidental (sunlight). Energy is more pervasive in the ways it enters production processes. Measurement is difficult and not always meaningful.

As a practical matter, decisions regarding energy use efficiency in irrigation relate primarily to the direct pumping energy used in delivering and applying water to crops. The indirect and embodied energy forms are generally of small consequence in a farm management setting. Direct pumping energy is frequently a major cost item that can be greatly influenced by irrigation management decisions.

Since substantial energy is often required to pump water to irrigated fields, the impacts of water use efficiency and energy use efficiency are closely linked. However, the association is not always one directional. Pursuing water conservation through increased water use efficiency may involve the movement from gravity flow to sprinkler irrigation systems and an increase in energy consumption. However, reducing water consumption in a setting that requires the use of high pump lifts will most likely lead to decreased energy consumption.

Economic criteria are at the center of decision making. Economic efficiency is, thus, invariably linked not just to water use or energy use efficiency but to the total inputs and outputs of the system. The farm firm is the level of most irrigation management decisions. Most farmers have objectives other than simply the maximization of farm profits. These may include such items as personal preference, leisure time, and the transfer of farm resources to the next generation. However, in most cases many of the goals of a well-managed firm can be expressed in terms of farm firm profits. Irrigation systems are complexes of physical attributes which operate in an economic and social environment. The choice of an efficiency measure should be appropriate for the decision being made.

REFERENCES

Batty, J.C., S.N. Hamad, and J. Keller. "Energy Inputs to Irrigation." Journal of the Irrigation and Drainage Division. American Society of Civil Engineers, IR4, pp. 293-307, December 1975.

Bernardo, Daniel and Norman K. Whittlesey. "Optimizing Irrigation Management With Limited Water Supplies." Selected paper, Western Agricultural Economics Association Meetings, 1985.

Bos, M.G. and J. Nugteren. "On Irrigation Efficiencies." Publication 19, International Institute for Land Reclamation and Improvement, The Netherlands, 1974.

Chervinka, V., W.J. Chancellor, R.J. Coffelt, R.G. Curley, J.B. Dobie, and B.D. Harrison. "Methods Used in Determining Energy Flows in California Agriculture." Transactions of the ASAE, pp. 246-251, 1975.

Christensen, R., H.T. Colwell, F. Turner, and J.C. Wade. "Evaluating Energy Management for Agriculture Within a Policy Framework: A Discussion Outline." In Energy Management and Agriculture, Proceedings first International Summer School in Agriculture, D.W. Robinson and R.C. Mollan, eds., Royal Dublin Society, Dublin, Ireland, pp. 35-42, 1982.

Dudek, D.J. and G.L. Horner. "The Derived Demand for Irrigation Scheduling Services." Selected paper presented at the annual meeting of the American Agricultural Economic Association, Urbana, Illinois, July 27-30, 1980.

Edwards, G.W. "Energy Budgeting: Joules or Dollars?"

Australian Journal of Agricultural Economics, 20(3): 179-191, 1976.

English, M.J., and G.T. Orlob. "Decision Theory Applications and Water Optimization." Contribution No. 174, California Water Resources Center, University of California, Davis, September 1978.

Erie, L.J., O.F. French, and K. Harris. "Consumptive Use of Water by Crops in Arizona." Technical Bulletin 169, Agricultural Experiment Station, The University of Arizona, Tucson, September 1965.

Farrell, M.J. "The Measurement of Productive Efficiency," Journal of the Royal Statistical Society. 120(Part III): 253-281, 1957.

Ferguson, C.E. The Neoclassical Theory of Production and Distribution. London: Cambridge University Press, 1971.

Fisher, A.C. Resource and Environmental Economics. London: Cambridge University Press, 1981.

Hagan, R.M., H.R. Haise, and T.W. Edminster, eds. Irrigation of Agricultural Lands. Number 11 in the Series AGRONOMY, American Society of Agronomy, Madison, Wisconsin, 1967.

Halderman, A.D. "Irrigation System Performance Criteria." Agricultural Engineering and Soil Science, Reprint 78-1, University of Arizona, Tucson, January 1978.

Harris, T.R. and H.P. Mapp, Jr. "Optimal Scheduling of Irrigation by Control Theory: Oklahoma Panhandle." Selected paper presented at the annual meeting of the American Agricultural Economic Association, Urbana, Illinois, July 27-30, 1980.

Hunt, V.D. Energy Dictionary. Van Nostrand Remholde Co., New York, 1979.

Michael, A.M. Irrigation Theory and Practice. Vikas Publishing House, PVT Ltd., New Delhi, 1978.

Israelsen, Orson W. and Hansen, Vaughn E. "Irrigation Principals and Practices." New York, John Wiley and Son, Inc., 1962.

Jensen, M.E., ed. Design and Operation of Farm Irrigation Systems. ASAE Monograph No. 3, American Society of Agricultural Engineers, St. Joseph, Missouri, 1980.

Ladd, G.W. "Value Judgements and Efficiency in Publicly Supported Research." Southern Journal of Agricultural Economics, 15(1): 1-8, 1983.

Peri, G., G.V. Skogerboe, and D. Karmal. "Procedures for Evaluation and Improvement of Irrigation Systems." Present Paper No. 79-2089, Joint Meeting of American

Society of Agricultural Engineers and Canadian Society of Agricultural Engineers, Winnipeg, Canada, June 24-27, 1979.

Spedding, C.R.W., J.M. Walsingham and A.M. Hoxey. Biological Efficiency in Agriculture. London: Academic Press, 1981.

Taylor, H.M., W.R. Jordon, and T.R. Sinclair, eds. Limitations to Efficient Water Use in Crop Production. American Society of Agronomy, Inc., Madison, Wisconsin, 1983.

U.S. Congress. Water-Related Technologies for Sustainable Agriculture in U.S. Arid/Semiarid Lands. Office of Technology Assessment, OTA-F-212, Washington, D.C., October 1983.

U.S. Department of Agriculture. "Irrigation Water Requirements." Technical Release No. 21, Soil Conservation Service, Engineering Division, Washington, D.C., April 1967.

Wade, J.C. "Conceptual Issues in Efficiency Measurement." Memograph, Department of Agricultural Economics, University of Arizona, Tucson, 1984.

Wade, J.C., R.J. McAniff and M. Flug. "Selecting Irrigation Systems Under Uncertain Energy Price Increases." Journal of the American Society of Farm Management and Rural Appraisal, 45(1): 23-30, 1981.

Warrick, A.W. and W.R. Gardner. "Crop Yield as Affected by Spatial Variations of Soil and Irrigation." Water Resources Research, 19(1): 181-186, 1983.

5
Concepts Affecting Irrigation Management

Norman K. Whittlesey, Brian L. McNeal, and Vincent F. Obersinner

INTRODUCTION

To increase the productivity of scarce water and expensive energy resources used in irrigated agriculture, it is first necessary to understand the complex and integrated relationships among crops, soils, and irrigation systems that can affect the outcome of irrigation management decisions. The following discussion provides an abstract view of the major factors that influence the effectiveness of irrigation management. While scientifically incomplete, it will provide the reader with an adequate understanding of irrigation to appreciate and evaluate the remainder of this book. The contents of this chapter are adapted from an earlier report by McNeal, Whittlesey, and Obersinner.

DETERMINATION OF IRRIGATION REQUIREMENTS

Irrigation Efficiency and Water Losses

Runoff and deep percolation water losses are a direct function of on-farm irrigation efficiency, which can be defined as the fraction of water applied to an irrigated field that is stored in the crop root zone. The remainder of the water is lost as evaporation, runoff, or deep percolation. In other words,

$$E = \frac{Q_a - Q_e - Q_r - Q_d}{Q_a} = \frac{Q_c}{Q_a}$$

where:

E = on-farm irrigation efficiency
Q_a = quantity of water applied
Q_e = quantity of water evaporated during application
Q_r = quantity of water which runs off
Q_d = quantity of water deep percolated
Q_c = quantity of water evapotranspired from the crop root zone (i.e., the net irrigation requirement)

It should be noted that this form of efficiency function differs from the overall system efficiency, which includes seepage losses from delivery canals, laterals, and on-farm ditches. Such seepage losses, which are operationally difficult to distinguish from deep percolation, can be extremely important in terms of project-wide water requirements, localized groundwater accumulation, and salinity pickup. In general terms, the runoff losses of irrigation water are largely responsible for soil erosion from irrigated lands and the associated phosphorous and nitrogen pickup losses. Deep percolation losses of water carry most of the nitrogen fertilizer lost from cropland and are the cause of nitrate accumulation in subsurface and surface receiving waters.

Values of Q_e usually range from 5 to 15 percent of Q_a for sprinkler-irrigated fields. A level of 10 percent is assumed for this discussion. Evaporative losses during application are assumed negligible for furrow-irrigated fields. Evaporative losses from the surfaces of recently-wetted (but empty) furrows are included in the Q_c values for a given crop and area. Values of Q_r are similarly assumed negligible for sprinkler-irrigated fields, though localized runoff can be substantial when water is applied at high rates to soils of low intake capacity. A common example can be observed near the periphery of center-pivot irrigation systems on fine-textured soils. Such runoff losses are assumed to remain primarily within-field, however. Values of Q_r are generally assumed to be more easily measured for furrow-irrigated fields than are values of Q_d, though this may not be true where well-designed subsurface drainage systems intercept virtually all deep percolation from an irrigated field without appreciable pickup of canal seepage or deep percolation from adjacent fields. In some cases, Q_d values are also estimated for sprinkler-irrigated fields as the difference between water applied and water calculated to be evapotranspired by the growing crop. Still another method of estimating or cross-checking Q_d values, from parameters related to the uniformity and adequacy of irrigation, will be discussed later in this section.

On-farm irrigation efficiency is a concept used widely throughout the irrigated West. Efficiency estimates are commonly available for the major crops and irrigation systems of a given physiographic region. Validity of the estimates range from carefully-researched numbers to values which may be little more than educated guesses. Reliable values will depend upon crop, irrigation system, slope, soil depth, soil texture, set time, length of run, and irrigation stream size.

Rooting Zone, Depletable Moisture, and Number of Irrigations

The crop root zone can be defined as the depth of soil which is penetrated by crop roots. It may or may not include a limited depth of soil from which water is removed by upward flow into the actual crop root zone. Typical rooting depths for common irrigated crops range from 2 feet for peas, beans, and potatoes to 4 feet or more for irrigated alfalfa, see Table 5.1 (USDA, 1973). Tabulated values generally assume that no obstructions to root growth are present, such as hard pans, chemically indurated layers, or dry soil.

Deep percolation is defined as that quantity of water which moves vertically through the soil below the crop root zone. Though subsequent upward movement is common under nonirrigated conditions, deep percolated water is normally unavailable for crop use in well-irrigated settings. The same would not be true where deficit irrigation is practiced.

Table 5.1
Typical rooting depths for major irrigated crops

Crop	Depth (ft)
Alfalfa	4
Small Grains	3
Pasture	3
Peas	2
Field Corn	3.5
Sugar Beets	3.5
Dry Beans	2
Potatoes	2

Source: Soil Conservation Service.

Coarse-textured soils allow more rapid infiltration and percolation of water than do fine-textured (high-clay) soils. For this reason, coarse-textured soils have a larger potential for deep percolation, and fine-textured soils have a larger potential for surface runoff and erosion. Soils that are extremely coarse-textured may even result in such large percolation losses that furrow irrigation is no longer practical, so sprinkler irrigation must be used.

Fine-textured soils hold more water than do coarse-textured soils. Typical water holding capacities for major soil texture classes range from one inch/foot for sands to four or more inches/foot for clay loams, Table 5.2 (USDA, 1973). These water holding capacities are derived from "field capacity" measurements, with field capacity defined as the amount of water remaining in a soil 24 to 48 hours after a heavy irrigation, under conditions of free drainage.

Depletable soil moisture can be defined as the quantity of moisture which may be withdrawn from the root zone without resultant crop moisture stress. At its maximum, it would be the water held between "field capacity" and the "permanent wilting point." At the permanent wilting point, however, soil water moves so slowly that it becomes insufficient to supply the daily crop water requirement. For

Table 5.2
Typical water holding capacities for major soil types

Soil Texture	Water Holding Capacity (in/ft)
Sand	1.08[a]
Sandy Loam	2.04[a]
Silt Loam and Loam	3.12[a]
Clay Loam	4.50[b]

[a]Source: Soil Conservation Service.
[b]Estimated by comparison with other values in the table.

sustained economic crop production under irrigated conditions, only about one-half of the total moisture range should be regarded as truly depletable. Unacceptable moisture stress would result if the wilting point were approached prior to each irrigation.

For a given crop, coarse-textured soils have a smaller depletable moisture range than do fine-textured soils. Crops with deeper root systems also have larger depletable moisture values, and crops vary in their sensitivity to moisture stress. For example, although most crops appear able to repeatedly remove up to approximately 50 percent of the depletable soil moisture without resultant yield losses, potatoes are often assumed able to regularly remove no more than 33 percent of the depletable soil moisture without a loss in yield or tuber quality.

Table 5.3 presents depletable soil moisture values for major crops and soil types of the Pacific Northwest. The values were computed using the relationship:

$$D = k(r)(w)$$

where:

D = depletable soil moisture in inches
k = percent of available soil moisture for plant consumption (equal to 0.5 for all crops except potatoes for which it is 0.33)

Table 5.3
Estimated depletable soil moisture values for major crops and soil types in the Pacific Northwest

Crop	Soil Type		
	Silt Loam or Loam	Sandy Loam	Sand
Alfalfa	6.24	4.08	2.16
Small Grains	4.68	3.06	1.62
Pasture	4.68	3.06	1.62
Peas	3.12	2.04	1.08
Field Corn	5.46	3.57	1.89
Sugar Beets	5.46	3.57	1.89
Dry Beans	3.12	2.04	1.08
Potatoes	2.08	1.36	0.72

r = crop rooting depth in feet
w = soil water holding capacity in inches per foot

Irrigation frequency and number of irrigations per season are determined by the time required for depletable moisture to be extracted from the crop root zone. Under ideal conditions, shallow-rooted crops or crops grown in coarse-textured soils would be irrigated more frequently than deep-rooted crops or crops grown in fine-textured soils. The relationship for determining numbers of irrigations (F) is:

$$F = Q_c/D$$

where:

Q_c = seasonal net irrigation requirement of the crop in inches
D = depletable soil moisture in inches

Table 5.4 shows estimated minimum numbers of irrigations for some representative crops based on the above criteria.

Table 5.4
Estimated numbers of irrigations per season for major crops and soil types in the Pacific Northwest

Crop	Silt Loam and Loam	Sandy Loam	Sand
Alfalfa	6	9	19
Spring Grain	5	7	a
Winter Grain	5	8	15
Pasture	5	8	a
Peas	5	8	a
Field Corn	4	6	15
Sugar Beets	5	8	a
Dry Beans	6	10	19
Early Potatoes	a	a	39
Late Potatoes	13	19	51

[a]Not estimated.

Some methods of irrigation such as center-pivot application may use many more irrigations to supply the same total crop needs for water.

In practice, factors such as water needs by other crops on the same farm or in the same area, water availability, needs of water for controlling wind erosion or for promoting selective crop or weed emergence, individual preference for light irrigations versus heavy irrigations, and availability of irrigation labor are also important in determining irrigation frequency. Although scientific irrigation scheduling is becoming increasingly more common, there remains considerable differences among growers as to when a given crop is judged ready for irrigation. At present, irrigation frequency generally remains a subjective determination.

Length of Set, Stream Size, and Length of Run

The time period that water is continuously applied to the field with furrow or sprinkler (not including center-pivot) irrigation can be defined as the length of set. Other factors being constant, the adequacy of irrigation should increase with length of set. Coarse-textured soils or shallow-rooted crops normally require a shorter length of set (but more frequent irrigations) than do fine-textured soils or deep-rooted crops.

In practice, the length of set is more commonly determined by convenience rather than by exact field needs. The most common lengths of set are 12 or 24 hours, so that sprinkler irrigation lines, or siphon tubes, can be changed at specified times each day. A less common length of set, for fine-textured soils, deep-rooted crops, or sprinkler lines which apply water at uncommonly low rates, is 48 hours. For center-pivot systems, care is usually taken to ensure that a single revolution is completed in <u>other</u> than a 12, 24, 36, or 48 hour period. This is done to prevent recurring application of water to a given part of the field at the same time of day, for less water remains in the root zone following mid-day irrigation than following irrigation at other times.

Stream size normally refers to the size of irrigation water stream delivered to the head of irrigation furrows. Because of infiltration along the furrow, it differs markedly from the size of stream flowing from the tail-end of that same furrow. Stream size is an important factor affecting furrow advance rate (the velocity at which the initial water stream moves down a previously dry furrow). A larger stream size results in a faster furrow advance rate at a given site, other factors remaining constant.

Other factors also affect furrow advance rate. Fields with steeper slopes, finer-textured soils, or relatively smooth furrow surfaces all result in greater furrow advance rates. Conversely, recently tilled furrows result in slower furrow advance rates. A shorter period of furrow advance to the end of the field results in less deep percolation, since the head of the field is then being irrigated for a shorter time in relation to the total length of set. Hence, the uniformity of application is increased. A faster furrow advance rate may also result in greater runoff, however. The optimal furrow advance rate is one which minimizes the sum of deep percolation and runoff losses. This is commonly established as the advance rate for which the total advance time (for the stream to initially reach the end of the furrow) is approximately one-fourth the length of set (e.g., five to seven hours for a 24 hour set). An alternative approach is to use cut-back irrigation, where a larger stream size is used to push the stream through to the end of the furrow in minimal time, and then the stream size is cut back to minimize runoff and erosion.

The length of an irrigation furrow from top to bottom of the field is termed the length of run. The shorter the length of run, the shorter the furrow advance time, other factors remaining constant. Use of a shorter length of run also allows the use of a smaller stream size, which implies a slower furrow advance rate and reduced runoff losses. Hence, the use of a shorter length of run can reduce both deep percolation and runoff, and increase irrigation efficiency. But the labor and management costs of irrigation can be increased by reducing length of run.

Coefficient of Uniformity and Adequacy of Irrigation

Any system of irrigation, whether it be furrow or sprinkler, results in nonuniform application of water over the field surface. In practice, for example, considerably more deep percolation occurs immediately beneath irrigation furrows than beneath adjacent crop rows, particularly on coarse-textured soils. In addition, the top portions of furrow-irrigated fields generally receive more water than do the bottom portions, producing greater deep percolation. Under sprinkler irrigation, overlapping sprinkler patterns result in double applications to some parts of the field, while other portions receive less than a full irrigation. Variations in irrigation application uniformity are often described by the "coefficient of uniformity," with a value of 100 percent denoting completely uniform water application. Values for existing sprinkler systems are commonly in the range 60 to 80 percent. Uniformity coefficients for furrow irrigated fields may be 50 to 60 percent or less. Improving the uniformity of water application results in a reduction in runoff and deep percolation losses, and increased irrigation efficiency.

The concept of "adequacy of irrigation" can be defined as the percentage of the root zone throughout a field which is restored to field capacity during each irrigation. Obtaining an adequacy level of 100 percent is generally not possible without incurring substantial percolation losses. In order to meet crop water needs, most fields are irrigated to an adequacy level of 75 to 87.5 percent. The latter level is commonly sought by U.S. Bureau of Reclamation design criteria, but is rarely met in fields for which variations in water application are normally distributed. The value of the crop, the importance of uniform field appearance to the grower, and the availability of water and nutrients all interact to determine the appropriate adequacy level for given irrigated settings.

Uniformity coefficient, adequacy level, and deep percolation (for normally-distributed variations in water application) are uniquely interrelated. If adequacy levels of 50 percent, 75 percent, and 87.5 percent are roughly equated to severely under-irrigated, "adequately" irrigated, and somewhat over-irrigated conditions, respectively, the interrelationship can be used as an alternative method of estimating deep percolation losses.

In assigning a proper adequacy level, careful consideration should be given to portions of the field which are never irrigated to mid-row because of convex field ends or inadequate set times. Consideration should be given as well to areas with reduced vegetative growth because of poorly-designed or wind-distorted sprinkler patterns.

SOME ALTERNATIVE IRRIGATION SYSTEMS

Some common irrigation systems will be briefly discussed in this section to illustrate the concepts of irrigation described above. The numerical data presented in the remainder of this chapter were derived from data pertinent to crop production in the Pacific Northwest. However, with proper attention to the environmental conditions underlying these data they are transferrable to many other irrigated areas. Many farmers do not know with any certainty how much water was being applied in a given irrigation. They typically apply water in 12 or 24-hour sets, often neglecting irrigating in order to perform more critical operations. Most irrigating is done on the basis of fixed rotation or visual crop symptoms, rather than by scheduling the irrigation to meet predicted crop demands. Farmers will frequently use a relatively small stream size in order to minimize visual runoff losses. As a consequence, they induce relatively large deep percolation losses because of nonuniform water application between head and tail portions of the field. Such existing furrow irrigation systems, for purposes of this discussion, will be designated by the acronym PFRW. Irrigation systems which have been improved via irrigation scheduling services, closer control of stream size, etc., will be designated as improved furrow (IFRW) systems.

Furrow irrigation efficiency can be improved still further by employing additional labor to reduce (cut back) stream size once the stream has advanced to the bottom of the field. In so doing, runoff losses can be reduced markedly. The initial use of a relatively large stream size also improves the uniformity of water application by reducing the period of furrow advance. Therefore, deep percolation losses are reduced as well. Cutback irrigation systems will be designated by the acronym CTBK.

With pump-back (PMPBK) furrow irrigation systems, runoff is collected from the bottom of one or more fields through the use of ponds, and then pumped back to the top of the same or adjacent fields for reuse. Although runoff still occurs, it does not leave the farm. The collection ponds also serve as sediment ponds, from which eroded sediment can be redistributed to nearby croplands. Deep percolation losses for pump-back systems have been assumed equivalent to those of present furrow irrigation systems. Some researchers have assumed reduced deep percolation losses from pump-back systems, reasoning that larger stream sizes can be employed since sediments are to be collected anyway. As pump-back systems are normally used on highly erosive fields, however, increased stream size would generally lead to excessive erosion and the subsequent need for unrealistically large collection ponds and frequent sediment spreading operations.

Gated pipe (GPIPE) irrigation systems combine features of both the improved furrow and the cutback systems. The gated pipe system can in turn be controlled by a time clock or a master control panel. When coupled with on-demand water availability, irrigations can be scheduled according to crop needs. Length of set can be automatically controlled, so that uniform set times are no longer required, and set length can even be varied during the course of the irrigation season. Stream size can be automatically cut-back when the stream reaches the end of the field.

The automatic multi-set (MLTST) irrigation system combines features of the improved furrow system with a shorter length of run, by having several lateral supply pipes across each irrigated field. Irrigations can be started automatically by time clocks or a master control panel. They can be scheduled to meet crop needs. Length of set can be automatically controlled, and the shorter length of run reduces the period of furrow advance. This in turn increases uniformity and decreases deep percolation. Additionally, a shorter length of run allows for the use of a smaller stream size, which reduces runoff.

Farmers using sprinkler irrigation systems (and especially center-pivot systems) frequently over-irrigate early in the season as intended insurance in case of a subsequent breakdown in the sprinkler system or a period when evaporative demand exceeds system capabilities. Increasing energy costs are causing farmers to reduce this form of over-irrigation. Early season irrigation is also used as a means of wind erosion control. Late season over-irrigation and leaching often result from a reluctance to decrease water application following the hot summer months as rapidly as plant water needs fall off. In view of these practices, and the fact that coarse soil textures predominate in areas where center-pivot irrigation is practiced, some deep percolation is common. Sprinkler irrigation systems considered for our studies include the side-roll (SDRL), current center-pivot (PCP), and improved center-pivot (ICP, implying an improved level of water management) systems. Solid-set, hand-move, permanent, big-gun, drip and/or trickle systems could be considered as well. They would be approached via similar techniques to those reported below.

Irrigation Efficiency Estimates

Furrow Systems. Sample estimates of furrow irrigation efficiencies for major crops and alternative irrigation systems are presented in Table 5.5. These values are expected to be representative of those now being attained in the Pacific Northwest.

Having an estimate of efficiency also gives an estimate of total water loss. The difficulty comes in dividing losses between runoff and deep percolation. Estimates of runoff are at least measurable (though often not conveniently), so values of runoff will commonly be estimated from existing data and deep percolation will be taken as the remainder of the total loss value. Runoff estimates of

Table 5.5
Estimated irrigation efficiencies for major crops and alternative irrigation systems in the Pacific Northwest

Crop	Irrigation System						
	PFRW	IFRW	CTBK	PMPBK	GPIPE	MLTST	SDRL
	---------------- percent ----------------						
Alfalfa	57.5	62.5	72.5	77.1	77.5	92.5	75.0
Small Grains	50.0	55.0	65.0	66.7	70.0	85.0	70.0
Pasture	50.0	55.0	65.0	67.6	70.0	85.0	70.0
Peas	47.5	52.5	62.5	62.7	67.5	82.5	62.5
Field Corn	45.0	50.0	60.0	69.2	65.0	90.0	72.5
Sugar Beets	45.0	50.0	60.0	69.2	65.0	90.0	72.5
Dry Beans	37.5	42.5	52.5	57.0	57.5	82.5	62.5
Potatoes	32.5	37.5	47.5	50.0	52.5	77.5	60.0

Note: Assumes silt loam and loam soils with 0.5 percent to 5 percent slopes.

25 percent and 35 percent are used herein for close-growing crops and row crops, respectively. Different values could be substituted for other areas, soil types, etc. Estimates of efficiency, runoff, or deep percolation are not refined to closer than ± 5 percent, because of the uncertainties of existing data.

Runoff estimates are deducted from irrigation efficiency estimates (expressed as percentages) to give estimates of deep percolation. Deep percolation should increase, however, as crop rooting depth decreases. Resultant runoff and deep percolation values for alternative crops and irrigation systems are given in Table 5.6.

Sprinkler Systems. Published efficiency estimates for side-roll irrigation systems generally range from 65 to 75 percent. Runoff losses from sprinkler-irrigated fields are assumed to be zero, and 10 percent of the applied water is estimated to be lost as spray evaporation and wind drift.

Table 5.6
Estimated percent runoff and deep percolation for major crops and alternative irrigation systems in the Pacific Northwest

Crop	PFRW	IFRW	CTBK	PMPBK	GPIPE	MLTST	SDRL
	Irrigation System						
	Runoff (percent)						
Row	35.0	32.5	25.0	0.0	22.5	5.0	0.0
Close-Growing	25.0	22.5	15.0	0.0	12.5	5.5	0.0
	Percolation (percent)						
Alfalfa	17.5	15.0	12.5	22.9	10.0	2.5	15.0
Small Grains	25.0	22.5	20.0	33.3	17.5	10.0	20.0
Pasture	25.0	22.5	20.0	33.3	17.5	10.0	20.0
Peas	27.5	25.0	22.5	37.3	20.0	12.5	27.5
Field Corn	20.0	17.5	15.0	30.8	12.5	5.0	17.5
Sugar Beets	20.0	17.5	15.0	30.8	12.5	5.0	17.5
Dry Beans	27.5	25.0	22.5	43.0	20.0	12.5	27.5
Potatoes	32.5	30.0	27.5	50.0	25.0	17.5	30.0

Note: Assumes silt loam and loam soils with 0.5 percent to 2 percent slopes. Percentages of loss relate to the quantity of water applied to meet full seasonal crop needs.

Deep percolation is estimated by deducting wind and evaporation losses from total losses for the sprinkler systems.

Irrigation efficiency values for center-pivot irrigation systems currently being used are not normally given for each crop. Differences among crops are generally assumed to be small and primarily reflect differences in rooting depth superimposed on the nonuniformity of sprinkler application. Such interaction leads to localized leaching in portions of the field receiving more water than the crop can use. The efficiency of sprinkler-irrigation systems is much more dependent on system design characteristics (as opposed to crop or soil characteristics) than is true for furrow-irrigation systems. The range in estimated efficiency values for center-pivot system is commonly from 75 to 85 percent and, for this discussion, an efficiency value of 80 percent is assigned. Field runoff is assumed to be zero for such systems (though considerable localized runoff can occur), and spray evaporation and wind losses assumed equal to 10 percent. The remainder of the losses (i.e., 10 percent) are attributed to deep percolation. A value of 5 percent is assumed as the improvement in efficiency to be realized as improved management techniques are applied to center-pivot systems (ICP). Similarly, deep percolation losses for improved center-pivot systems are assumed to equal 5 percent. Spray evaporation and wind losses remain unchanged.

Effect of Texture and Slope on Irrigation Efficiency

Since various soil textures exist throughout the irrigated West, it is necessary to have some feeling for the effects of texture on efficiency, runoff, and deep percolation. It is frequently assumed that furrow irrigation efficiencies show negative adjustments (i.e., greater losses) for more coarse-textured soils. However, runoff losses decrease for coarse-textured soils, which tends to offset an increase in deep percolation. Since these authors feel that the increase in deep percolation tends to be greater than the decrease in runoff as soils become more coarse-textured, a decrease in efficiency of 5 percent is used as the dominant soil texture is changed from silt loam to sandy loam.

For sprinkler irrigation systems, it is also believed that reductions in efficiency are required as soils become more coarse-textured. This analysis assumes a reduction in efficiency of 5 percent as the dominant soil texture is changed from silt loam to sandy loam.

Hence, all efficiency values for base-level (silt loam) soils are reduced by 5 percent in order to obtain corresponding irrigation efficiency values for sandy loams. The deep percolation values for silt loams or loams are also adjusted upward by 5 percent in order to obtain deep percolation values for sandy loams. No soil-texture-dependent adjustment of runoff losses is required by this methodology, though the sandy loam soils are assumed to be less erosive per unit of runoff volume than their silt loam counterparts.

It is also important to understand the effects of field slope on irrigation efficiency, runoff, and deep percolation. Published values suggest that efficiency of both furrow and sprinkler systems generally decreases as slope increases, though sprinkler system estimates often don't vary until after the slope has been increased to five or 10 percent.

Control of furrow stream size is more difficult on steeper slopes, so the percent runoff tends to increase. The stream also tends to run down the furrow more rapidly on steeper slopes, however, resulting in a smaller wetted cross section. This should operate to decrease deep percolation, and may in some cases approximately balance the increased runoff. For this discussion, furrow irrigation efficiency is assumed to remain unchanged in going from a 0.5 to 2 percent slope class to a 2 to 5 percent slope class. No adjustment in efficiency is generally required for sprinkler irrigation systems during this particular change in slope class as well. However, the runoff and deep percolation values for the base-level (0 to 2 percent slope class) conditions are respectively increased by 5 percent and decreased by 5 percent for furrow irrigation systems on slopes of 2 to 5 percent.

NET IRRIGATION REQUIREMENTS

Crop consumptive use is defined as the amount of water transpired during plant growth plus that evaporated from the soil surface and foliage of the area occupied by the growing plant (Sutter and Cory). The net irrigation requirement is commonly defined to be crop consumptive use less seasonal precipitation (Watts, et al.). It does not consider moisture stored in the root zone at the beginning of the growing season. As a result, estimates of net irrigation requirement somewhat overstate seasonal crop needs for irrigation water, especially if "deficit" irrigation (using previously stored soil moisture) is to be practiced. Despite this, net irrigation requirement estimates are commonly used, because other appropriate water use data are limited.

The net irrigation requirement of a crop is divided by the assumed irrigation efficiency to determine the necessary water application requirement. This is the quantity of water applied to the field through irrigation.

Once the quantity of water to be applied and of the percent runoff and deep percolation are determined, it is then possible to obtain estimates of the amount of runoff and the amount of deep percolation. Water application requirements for major crops and alternative irrigation systems in the Pacific Northwest are presented in Tables 5.7 and 5.8.

ESTIMATING DEEP PERCOLATION LOSSES

Several methods can be employed to estimate deep percolation losses. The approach recommended here consists of estimating deep percolation (as a percentage of water application to a given field) as the on-farm irrigation efficiency (expressed as a percentage) minus the percent runoff.

For some sprinkler-irrigated systems, deep percolation estimates can be obtained from estimates of uniformity coefficient and adequacy level (if such are available). Water distributions for three uniformity coefficient values are shown in Figure 5.1, assuming a normal distribution of variations in water application. A field net irrigation requirement of 1 inch and a 50 percent adequacy level is shown, with only one-half of the root zone soil being returned to its maximum water-holding capacity during irrigation. Some subunits of the field (shown as 10 percent increments or 0.1-field subunits) must be over-irrigated even to meet this adequacy level. As a result, deep percolation occurs. Differences in deep percolation for the various uniformity coefficients are obvious. The tradeoff between under-irrigation and deep percolation is also apparent from the distributions. As less and less of the root zone of a field is under-irrigated, by increasing the amount of water applied to each subunit of the field (refer to Figure 5.2), an increasing area is subjected to deep percolation.

As our knowledge of irrigation water distribution and soil water movement improves, the adequacy level may eventually be amenable to fairly precise estimation. It cannot be measured or even precisely estimated with current technology, however. Despite such difficulties, the adequacy level can at least be qualitatively estimated. The adequacy levels considered here are 50 percent, 75

Table 5.7
Estimated seasonal water application requirements for major crops and alternative irrigation systems in the Pacific Northwest

Crop	Irrigation System						
	PFRW	IFRW	CTBK	PMPBK	GPIPE	MLTST	SDRL
	------------- inches/acre --------------						
Alfalfa	64	59	51	48	48	40	49
Spring Grains	44	40	34	33	32	26	32
Winter Grain	50	46	39	37	36	30	36
Pasture	50	46	39	37	36	30	36
Peas	34	31	26	25	24	20	26
Field Corn	49	44	37	32	34	24	30
Sugar Beets	64	57	48	42	44	32	40
Dry Beans	54	47	38	35	35	24	32
Late Potatoes	80	69	55	52	50	33	43

Note: Assumes silt loam and loam soils with 0.5 percent to 5 percent slopes.

percent and 87.5 percent. These can be roughly equated to severely under-irrigated, "adequately" irrigated, and somewhat over-irrigated, respectively.

Table 5.9 demonstrates the interrelationship of adequacy of irrigation, uniformity coefficient, and deep percolation values for normally-distributed water application systems. Fixing any two of these parameters automatically fixes the third. For example, uniformity coefficient estimates are commonly available for sprinkler irrigation systems. Hence, estimation of deep percolation losses is straight-forward once an adequacy level has been assigned from the descriptive terms: under, adequately, or somewhat over-irrigated. If, on the other hand, uniformity coefficient and on-farm irrigation efficiency values are already known, an appropriate value (often 5 to 10 percent) can be assigned to wind and evaporation losses.

Table 5.8
Estimated water application requirements for major crops and the center-pivot irrigation systems in the Pacific Northwest

Crop	Irrigation System	
	PCP	ICP
	------ inches/acre -------	
Alfalfa	51	48
Winter Grain	31	30
Field Corn	36	34
Dry Beans	26	24
Early Potatoes	35	33
Late Potatoes	46	44

Note: Assumes sandy soils.

Subtraction of this value plus the percent irrigation efficiency from 100 yields the percent deep percolation. Comparing this value with the corresponding deep percolation value from Table 5.9 provides an independent check on the adequacy level which should have been assigned to this field.

For furrow-irrigation systems, the uniformity coefficient will commonly be unknown, though often in the range 60 to 80 percent. Estimates of on-farm irrigation efficiency (often 40 to 60 percent for furrow-irrigated fields) can be used to estimate deep percolation losses, after subtracting the percent runoff and the percent efficiency from 100. If the resultant deep percolation estimate corresponds to an unrealistic uniformity coefficient or adequacy level (Table 5.9), a different on-farm efficiency or percent runoff may be required to "fine tune" the deep percolation, uniformity coefficient, and adequacy level values.

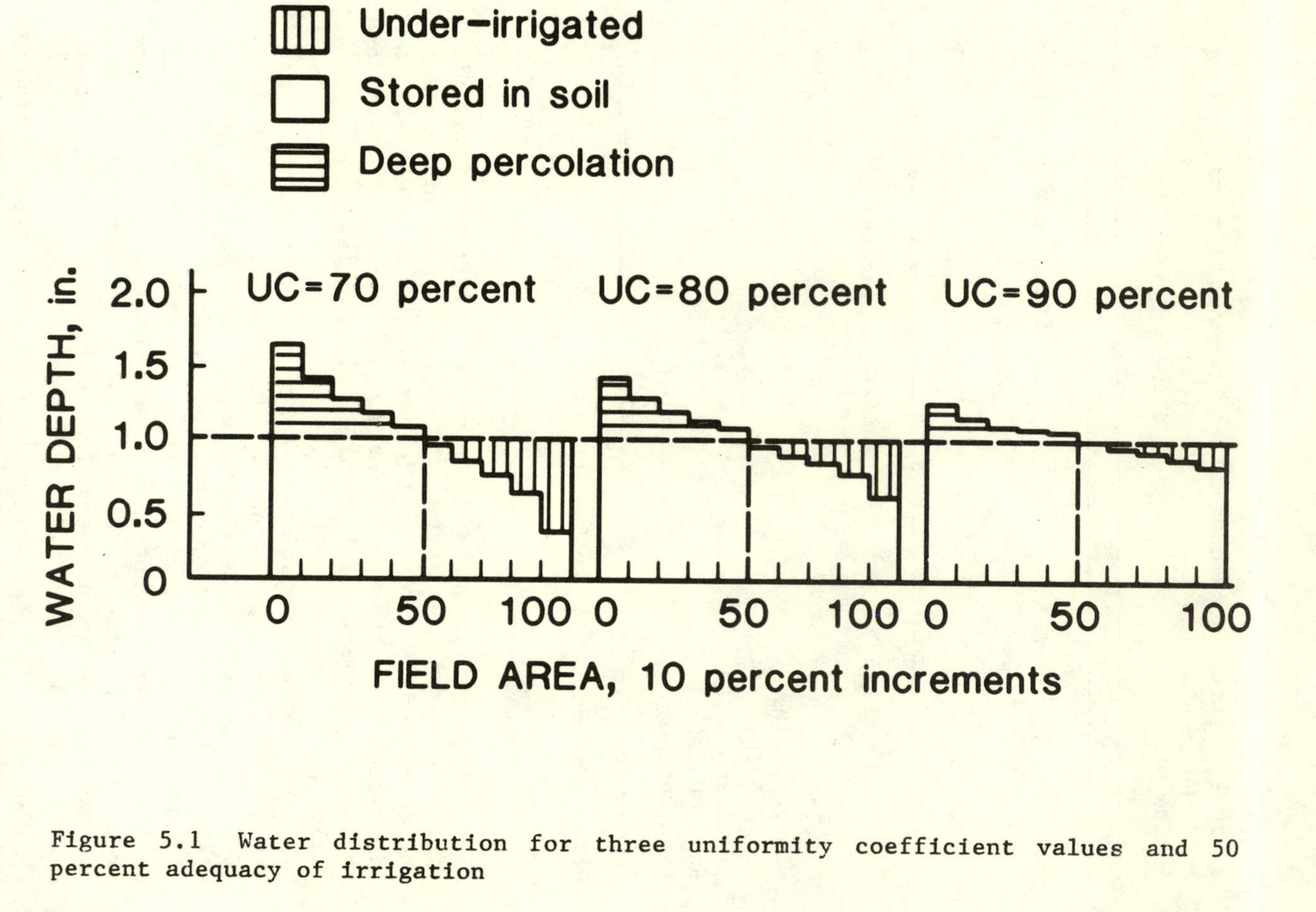

Figure 5.1 Water distribution for three uniformity coefficient values and 50 percent adequacy of irrigation

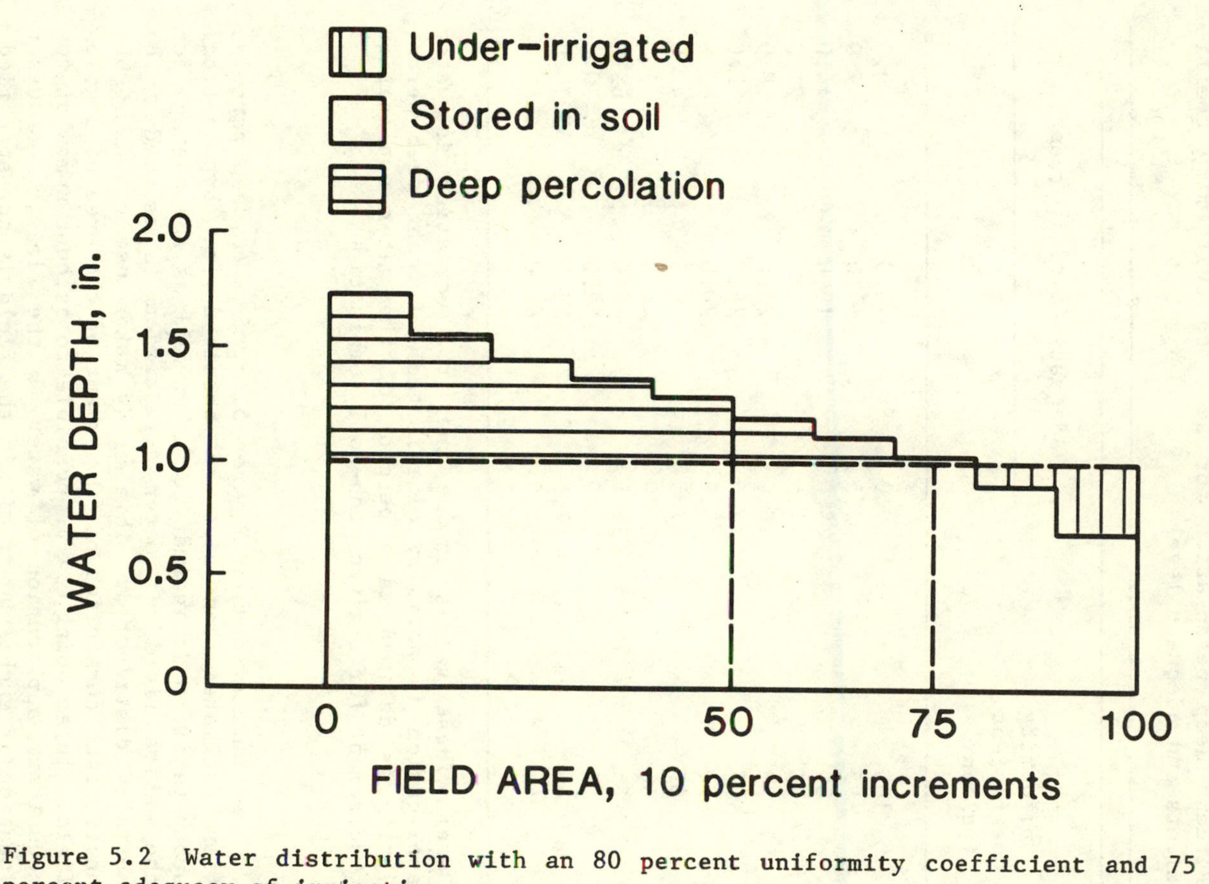

Figure 5.2 Water distribution with an 80 percent uniformity coefficient and 75 percent adequacy of irrigation

Table 5.9
Percent deep percolation for selected uniformity coefficients and adequacy levels

Uniformity Coefficient (percent)	Adequacy Level (percent)		
	50	75	87.5
100	0.0	0.0	0.0
95	2.5	5.0	7.0
90	5.0	10.0	14.0
85	7.5	15.5	21.0
80	10.0	21.0	28.0
75	12.5	26.0	35.0
70	15.0	30.0	42.0
65	17.5	36.0	49.0
60	20.0	41.0	56.0
55	22.5	46.0	63.0
50	25.0	52.0	70.0
45	27.5	58.0	77.0
40	30.0	62.0	84.0

Note: These values assume that the variation in water application is normally distributed. Percent deep percolation is defined as a percent of the water application requirement for a given crop-soil combination.

As suggested from Figures 5.1 and 5.2, deep percolation estimates can also be obtained for hypothetical subunits of an irrigated field, as well as the average for the entire field, if water application is assumed to be normally distributed. This study makes use of such subunits, each consisting of 10 percent of the area of a given field. In so doing, it is possible to account somewhat for nonuniform application of water to the field. The exact location of each subunit within the field is not specified, but neither is it necessary. By assigning an adequacy level and uniformity coefficient, and assuming water application to be normally distributed, it is possible to estimate the amount of deep percolation or under-irrigation for each field subunit (Figures 5.1 and 5.2). Table 5.10 shows calculated values of deep percolation for various uniformity coefficients at three adequacy levels. Deep percolation is expressed as a percent of the average amount of water applied to the field. As a result, it is possible to have deep percolation percentages of greater than 100 percent for a given subunit. Focusing for a moment on the 75 percent adequacy level, one can see that deep percolation does not occur for subunits eight through ten. Hence, these subunits are located in parts of the field that are under-irrigated.

For furrow-irrigated fields, ridges of the crop row and the lower part of the field (especially where convex field ends occur because of prior erosion) should correspond to field subunits that are not adequately irrigated. Irrigation furrows and the upper part of the field, on the other hand, should correspond in general to subunits that are over-irrigated. For sprinkler-irrigated fields, the water application pattern is not uniform due to design problems and/or wind distortion of the designed application pattern.

CONCLUSIONS

Understanding the concepts of irrigation described in this chapter is important to those persons studying or otherwise desiring to affect the management and productivity of water and energy resources used in irrigated agriculture. While this discussion has been terse and nonglobal, the reader is provided a basic understanding of the major factors that influence irrigation efficiency. It is clear that the physical and environmental system associated with an irrigated regime is very complex. Precise estimates of parameters affecting the system operation are difficult to obtain and are highly interrelated. It is difficult to change one parameter or factor without influencing many others in the system.

A major use of the water loss estimates from irrigation is to provide a means for determining the environmental effects of irrigated agriculture. Erosion and sediment losses from irrigated lands are related to total water runoff, in addition to such factors as type of crop, soil type, and method of irrigation. Phosphate nutrient

Table 5.10
Percent deep percolation for tenth-field subunits at selected uniformity coefficients and adequacy levels

Uniformity Coefficient (percent)	Field Subunit[a]									
	1	2	3	4	5	6	7	8	9	10
	----------- 50 percent adequacy level -----------									
100	--	--	--	--	--	--	--	--	--	--
95	10	6	4	2	1	--	--	--	--	--
90	21	13	9	5	2	--	--	--	--	--
85	32	20	13	8	2	--	--	--	--	--
80	43	25	17	10	3	--	--	--	--	--
75	54	32	21	12	4	--	--	--	--	--
70	65	39	25	15	5	--	--	--	--	--
65	76	46	30	17	6	--	--	--	--	--
60	87	52	34	19	6	--	--	--	--	--
55	98	59	38	22	7	--	--	--	--	--
50	110	66	43	24	8	--	--	--	--	--
45	121	72	47	26	9	--	--	--	--	--
40	132	79	51	29	10	--	--	--	--	--
	------------ 75 percent adequacy level -----------									
100	--	--	--	--	--	--	--	--	--	--
95	15	11	8	6	5	4	2	--	--	--
90	30	22	17	13	10	7	4	--	--	--
85	45	32	26	20	15	10	6	--	--	--
80	60	43	34	27	20	14	7	--	--	--
75	76	54	42	34	25	18	9	--	--	--
70	91	65	51	40	30	21	11	--	--	--
65	106	76	60	46	35	24	13	--	--	--
60	121	86	68	53	40	28	15	--	--	--
55	137	97	76	60	45	32	16	--	--	--
50	153	108	85	67	50	35	18	--	--	--
45	168	119	94	74	55	38	20	--	--	--
40	183	130	102	80	60	41	22	--	--	--

(Continued)

Table 5.10 (Cont.)

Uniformity Coefficient (percent)	Field Subunit[a]									
	1	2	3	4	5	6	7	8	9	10
---------- 87.5 percent adequacy level ----------										
100	--	--	--	--	--	--	--	--	--	--
95	17	13	11	9	8	6	4	2	--	--
90	34	26	22	18	15	12	8	5	--	--
85	51	39	32	27	22	18	12	7	--	--
80	69	52	43	36	29	23	17	9	--	--
75	86	66	54	45	36	29	21	12	--	--
70	104	79	65	54	44	35	25	14	--	--
65	122	94	76	63	52	40	29	16	--	--
60	139	105	86	72	59	46	33	19	--	--
55	157	118	97	81	66	52	37	21	--	--
50	175	131	108	90	73	58	41	23	--	--
45	193	144	119	99	80	64	46	26	--	--
40	211	157	130	108	88	69	50	28	--	--

[a]Each field subunit is defined as 10 percent of the area of the field. Deep percolation for subunits nine and ten equals zero for all uniformity coefficients and adequacy levels listed.

Note: Percent deep percolation for each tenth-field subunit is defined as a percentage of the average water application requirement for the field.

losses normally occur through the soil erosion process. Similarly, the nitrate-nitrogen that finds its way into ground and surface waters is largely a function of deep percolation losses. It can be seen how irrigation management must proceed to affect these environmental concerns. However, raising irrigation water use efficiency to reduce these losses is often accompanied by increases in energy use and increased capital requirements. Farmers are very sensitive to these tradeoffs.

REFERENCES

Gardner, W. and C.W. Lauritzen. "Erosion as a Function of the Size of the Irrigation Stream and the Slope of the Eroding Surface." Soil Sci., 62: 233-242, 1946.

Gossett, D.L. "The Economics of Changing the Water Quality of Irrigation Return Flow from Farms in Central Washington." M.S. thesis, Washington State University, Pullman, 136 pp., 1975.

Gossett, D.L. and N.K. Whittlesey. "Cost of Reducing Sediment and Nitrogen Outflows from Irrigated Farms in Central Washington." Washington Agricultural Experiment Station Bulletin 842, Washington State University, Pullman, 136 pp., 1975.

Hagan, R.J. "Energy in Western Agriculture--Requirements, Adjustments, and Alternatives." California Contributing Project Report, Western Regional Research Project, Department of Land, Air and Water Resources, Water Science and Engineering Section, Davis, California, 1978.

Humphreys, A.S. "Automatic Furrow Irrigation Systems." Transactions of the ASAE, 14: 466-470, 1971.

Jensen, M.E., J.L. Wright, and B.J. Pratt. "Estimating Soil Moisture Depletion from Climate, Crop and Soil Data." Transactions of the ASAE, 14: 954-959, 1971.

McNeal, B.L., N.K. Whittlesey, and L.G. King. "Factors Influencing Sediment Loss from Some Pacific Northwest Irrigated Croplands." Agronomy Abstracts, American Society of Agronomy, Madison, Wisconsin, p. 207, 1979.

McNeal, B.L., N.K. Whittlesey, and V.F. Obersinner. "Controlling Sediment and Nutrient Losses From Pacific Northwest Irrigated Areas." EPA 600/2-81-090, 1981.

Obersinner, V.F. "The Economics of Reducing Pollution from Irrigation Return Flows in Selected Areas of the Pacific Northwest." M.S. thesis, Washington State University, Pullman, 231 pp., 1979.

Pfeiffer, G.H. and N.K. Whittlesey. "Economic Costs of Controlling Water Quality through Management of Irrigation Return Flows." State of Washington Water Research Center Report No. 28, Pullman, 58 pp., 1976.

Pope, D.L. and A.D. Barefoot. "Reuse of Surface Runoff from Furrow Irrigation." Transactions of the ASAE, 16: 1088-1091, 1973.

Rasmussen, W.W., J.A. Bondurant, and R.D. Berg. "Multiset Surface Irrigation System." International Committee on Irrigation and Drainage Bulletin, pp. 48-52, 1973.

Shearer, M.N. "Comparative Efficiency of Irrigation Systems." Irrigation Association Technical Conference, Cincinnati, Ohio, 6 pp., 1978.

Soil Conservation Service, U.S. Department of Agriculture. "Irrigation Guide for Columbia Basin." Spokane, Washington, 1973.

Sutter, R.J. and G.L. Cory. "Consumptive Irrigation Requirements for Crops in Idaho." Agricultural Prices, Annual summaries for 1975, 1976, and 1977, Washington, D.C., 1975-1977.

Vomocil, J.A. "Estimated Irrigation and Water Application Requirements: Northern Morrow County and Northwest Umatilla County, Oregon." Project Working Paper No. 1, Northern Columbia River Basin Irrigation System Development Project, Pacific Northwest Regional Commission, Oregon State Univ., Corvallis, 1976.

Worstell, R.V. "An Experimental Buried Multiset Irrigation System." American Society of Agricultural Engineers Paper No. 75-2540, Chicago, 1975.

Worstell, R.V. "Costs and Benefits of a High Efficiency, Automatic Gravity Irrigation System." Pacific Northwest Regional Meeting of American Society of Agricultural Engineers, Penticton, B.C., Canada, 1976.

6
Management Systems for Improving On-Farm Energy and Water Use Efficiency

Wyatte L. Harman

INTRODUCTION

Improvements in energy use efficiency have been emphasized since the OPEC oil embargo of 1973. Energy intensive irrigation practices common to western agriculture were suddenly threatened by the potential limited availability and sharply rising price prospects of energy. Many production areas in the arid and semi-arid West which rely on groundwater with extensive pumping lifts were particularly vulnerable.

A recent survey of irrigation practices and estimated irrigation costs based on 1978 data indicated approximately 70 percent of the irrigation water from underground sources cost over $20 per acre-foot in the western U.S. This compared to about 20 percent at this cost level for surface water sources (Wilson and Ayer).

Gilley, Heerman, and Stetson point out that:

> "The largest potential savings of energy use in irrigation can be obtained by reducing the quantity of water pumped, and any reduction in water pumped will yield a corresponding reduction in energy consumption, even if no other changes are made in the irrigation system or pumping plant."

Alternative management practices suggested by Gilley, Heerman, and Stetson to save energy or reduce energy costs include:

1. reduction of the net water applied,

2. improvement in the irrigation efficiency (defined as the fraction of the water stored in the root zone of that pumped),
3. reduction in the total dynamic head (includes pumping lift, pumping pressure, and friction losses),
4. improvement in pumping plant performance,
5. decreasing peak electrical demands by off-peak irrigation scheduling,
6. improved irrigation management to prevent deep-percolation, and
7. adoption of conservation tillage practices.

The above suggestions are proposed for reducing the direct energy requirements for irrigating crops. Other energy requirements in the production process include associated inputs such as tractor fuel, chemicals, and the embodied energy requirements for machinery and equipment manufacture, transportation, and marketing. Patrick and Hobson found that in areas of groundwater pumping in New Mexico and Colorado, the direct energy requirements compose the largest share of total energy needs in crop production. For example, in New Mexico, direct energy requirements for irrigating wheat amount to 63 percent of the total; cotton, 87 percent; peanuts, 69 percent; alfalfa, 73 percent; and irrigated sorghum, 61 percent.

In California where water sources include surface, groundwater, and combinations of the two, Roberts and Hagan estimated that 73 percent of the total energy used for irrigation in 1972 was for providing water to the farm, 15 percent for pressurizing systems and nearly 12 percent was associated with the manufacture, installation and maintenance of application systems including land preparation for surface irrigation. Larson and Fangmeier found that in Arizona total direct and indirect energy requirements are lowest when using surface water and surface irrigation. Adding a sprinkler system for distribution raises the energy requirements by 5 to 13 fold depending on the crop requirements. Using groundwater as a source requires over 25 times more energy than when surface water is the source. There are, however, some areas such as the Pacific Northwest where surface water is also lifted several hundred feet for irrigation (Whittlesey).

Economic implications of the above studies indicate that management systems increasing energy use efficiency are most urgently needed where groundwater is the source of irrigation, where surface water irrigation is accompanied by high lifts, or where pumping costs are escalating rapidly. The following discussion describes research results attained by improved irrigation practices, peak load scheduling, reduced tillage cultural practices, crop rotations, land forming techniques, and other related management alternatives being developed by the scientific community.

ALTERNATIVE MANAGEMENT SYSTEMS FOR MORE EFFICIENT ENERGY USE

Several alternative irrigation and crop production management practices useful in reducing energy utilization will be discussed. They will be categorized into four major areas: (1) improved irrigation practices, (2) land forming techniques, (3) minimum tillage and innovative crop rotations, and (4) miscellaneous management options.

Improved Irrigation Practices

Optimum Timing and Application Rates. High on the priority of increasing energy use efficiency is the optimum timing and rate of irrigation applications. Scheduling irrigations at the proper frequency, rate, and duration during peak crop water use periods can increase production per unit of input and thus, enhance energy use efficiency. Stegman, Musick, and Stewart indicate many applied experiments have shown that yields of determinate crops (crops which produce a uniformly maturing seed set) are most sensitive to water stress in the initial floral through pollination stage of growth. Yield is typically least sensitive to mild stress in the early vegetative stage for this type of crop. Yield sensitivity also tends to diminish upon grain filling and maturity. Robins and Domingo; and Stewart, et al. (1975) support these conclusions for corn.

Plant stress can be quantified by the use of a stress day index developed by Hiler and Clark. The index provides a means of determining the relation between water stress and yields of crops. Shipley and Regier found that if only one irrigation is applied to sorghum in the Texas High

Plains in addition to a preplant irrigation, the highest energy use efficiency is attained with irrigation applied at boot stage of growth, Table 6.1. Jensen and Musick, Musick (1968), and Musick and Stewart found similar results. If two additional irrigations are to be applied, Shipley and Regier found that the highest energy use efficiency was achieved with irrigation at the boot and heading stages of growth (Table 6.1). Similarly, with three irrigations (not shown), the highest energy use efficiency was attained with irrigations at the six- to eight-leaf stage, boot and heading. Shipley went on to evaluate the optimum timing of corn irrigations in the Texas Panhandle. Optimum

Table 6.1
Energy use efficiency of grain sorghum irrigations, North Plains Research Field, Etter, Texas, 1969

Number of Irrigations[a]	Yield (lb/ac)	Energy Use Efficiency[b] (lb/mcf)
PP + E	799	137
PP + B	4,019	687
PP + H	3,167	541
PP + M	1,141	195
PP + E,B	3,659	433
PP + E,H	4,181	495
PP + E,M	1,260	149
PP + B,H	5,237	620
PP + B,M	3,677	435
PP + H,M	3,954	468

Source: Adapted from Shipley and Regier. Multiply lb/ac by 1.120851 to get kg/ha.

[a] 4-inch seasonal irrigations, 5-inch preplant irrigation, E - 6-8 leaf, B - boot, H - heading, M - milk.
[b] .65 mcf natural gas required per inch of water.

times for irrigating corn were determined to be in the respective order of tasseling, pre-tassel, blister, soft dough, and last, that of the hard dough stage of growth.

Schneider, Musick, and Dusek evaluated winter wheat irrigation response in 1966 and 1967 in Texas. One well-timed spring irrigation increased yields more than two poorly timed applications. The most sensitive period was found to be from boot through early grain filling. These findings agree with Robins and Domingo (1962) in Washington.

Many commercial crops grown in the West, such as wheat sorghum, corn, cotton, and many seed crops can be irrigated with improved timing and application rates to achieve both increased water and energy use efficiency. Advances in computer technology, soil moisture monitors, water budgeting techniques, remote sensing, improved weather forecasting techniques, and growth in management services for producers will improve timing of irrigation applications on a wider scale.

Peak Load Scheduling. Rapidly rising energy costs since the mid 1970s have been instrumental in the development of irrigation schemes such as peak load scheduling to reduce energy costs (Stetson, et al.). Electricity suppliers sometimes charge higher rates during peak demand periods since their cost is based on providing generating facilities sufficient to meet the highest demands. Scheduling irrigations during off-peak power demand periods can reduce the total cost of producing that power and, hence, the cost charged to the irrigator. Alternatives proposed by Heerman and Duke (1978); and Stetson, et al. include (1) voluntary reduction in pumping during peak demand periods, (2) controlled interruption of pumping units, (3) water management integrated with load management, (4) redesign of the system to reduce pressure and discharge, and (5) increasing the pumping plant efficiency. The latter two options serve only to decrease overall energy demands during the growing season as opposed to reducing energy requirements during a specific shorter peak demand period.

Power load management will reduce peak demand requirements but Heerman and Duke point out that it may result in more water applied by producers to insure adequate soil moisture availability for the crop. Planned use of water stored in the root zone is the key to off-peak scheduling of irrigations.

Buchleiter, Heerman, and Duke simulated power demands of a substation and found that by combining water management with load management the peak demand could be reduced up to 20 percent with no loss of crop yields. An experiment in California with customers of two utilities (California Public Utilities Commission) found that average peak-period load reductions would amount to 10 to 13 percent if time-of-use meters were installed and daily peak-period time-of-use rates were initiated.

Alternate Furrow Irrigation. Irrigation of alternate furrows has been evaluated as a means of reducing irrigation water applications where row irrigation is used. Musick and Dusek, using furrow spacing of 30 and 40 inches and irrigating alternate rows, found reduced water intake of 17 to 33 percent by sorghum and sugarbeets on the slowly permeable Pullman clay loam soil type in the Texas High Plains. The major reduction in water intake occurred on the lower one-fourth to one-half of the field. Sorghum yields dropped by over 200 pounds per acre using 30-inch furrows and by nearly 800 pounds or 11 percent on 40-inch rows. Neither less water nor less yields were experienced on the more permeable Pullman silty clay loam soil type.

Fischbach and Mulliner indicated that alternate furrow irrigation of 30-inch corn on several soil types in eastern Nebraska produced yields similar to irrigating every furrow. In this situation, irrigating alternate rows on Sharpsburg silty clay loam resulted in 29 percent less water use while yields fell by only 4.7 percent. Grimes, Walhood, and Dickens found that irrigating alternate furrows of cotton in the San Joaquin Valley on Hisperia sandy loam reduced the irrigation requirements by 23 percent while lint yields were not reduced at all. Yields of potatoes have also been successfully maintained using this approach (Box, et al.). Alternate furrow irrigation practices apparently can conserve energy in certain situations such as medium to coarse-textured soils or where relatively narrow rows are utilized.

Reduction in Row Length. Reducing the length of run with furrow irrigation to prevent deep percolation losses was evaluated in 1961-1965 (Musick, Sletten, and Dusek). Slowly permeable soils such as the Pullman clay loam in Texas are not thought to be subject to large deep percolation losses. However, water intake rates were found to decrease with shortened runs due to the reduced flow duration time but yields of sorghum also dropped. Water use efficiency was, however, increased and thus, a higher energy use efficiency resulted. To prevent yield losses from occurring on slowly permeable soils, flow duration time must be maintained to prevent reducing water intakes. Water runoff, however, will vary and will depend on soil permeability and land slope. Consequently, tailwater reuse and "cutback" irrigation practices have been suggested to prevent excessive water runoff losses (Criddle, et al.).

Recirculation of Irrigation Runoff. Recirculation systems or tailwater reuse systems are primarily designed to recapture water runoff incurred from the extended flow duration time needed to wet the length of furrow uniformly. Rainfall runoff can also be captured. Fischbach and Somerholder increased average irrigation uniformity from 65 percent to 92 percent with a reuse system.

In the Texas High Plains, many producers have already recognized the importance of the potential yield losses and reduced profits associated with poor uniformity and reduced water intake on the lower end of the irrigation run. To prevent wasting the water runoff when extended flow duration time is required to adequately wet the complete length of the furrow, tailwater, or recirculation pumps are installed in preconstructed pits or in natural playa lakes.

These measures, however, require additional investment capital for pumps, power plants, and distribution lines. In addition, the practice involves some repumping costs of the tailwater which has already incurred the original pumping cost. However, an economic analysis of this conservation measure by Young found that the present worth of increased profits ranged from $42 to $52 per acre where zero to 1 percent slopes were assumed to have 10 percent runoff and from $141 to $191 per acre where 30 percent runoff occurred on 1 to 3 percent slopes.

"Cutback" Irrigation Systems. Irrigation "cutback" systems, if manually controlled, are labor consuming and pose a problem for handling excess water during the reduced-flow period. Thus, a completely automated surface distribution system would be an improvement. Automation of surface distribution systems is difficult since water must be uniformly distributed to each furrow. Border or basin irrigation is more suited to automation. One technique to achieve automatic cutback streams is to utilize a storage pond for providing the high flow rate as a supplement to a constant but low flow (Fischbach). Humphreys suggests a split-set technique where the first half of the set is irrigated at full flow until water runs off the field and then apply a full flow to the other half set for the same time period, and finally, spread the water over the entire set at the reduced flow rate. This is a similar concept to the "surge flow" system discussed by Stringham and Keller.

Lined ditch or canal cutback systems can be achieved simply by constructing a ditch with a series of level bays with equally elevated outlets in each. As water progresses from bay to bay via timed check gates, the level is lowered in the upper bays automatically reducing the flow from the outlets (Garton; Nicolaescu and Kruse; Hart and Borrelli; Evans). A recirculation pit will probably be required with most cutback systems since water advances at varying rates down furrows due to differences in soil compaction, soil moisture, residue, and tilth.

The energy and water savings of a "cutback" system will vary depending on soil type, slope and amount of water applied. Savings of deeply percolated irrigation water result in reduced energy for pumping. The method is particularly useful on coarse-textured soils (high intake rates) with modest slopes where deep percolation of water and plant nutrients below the root zone results in higher pumping costs and increased fertilizer needs.

Land Forming Techniques

Furrow Diking. A technique developed in 1931 by C.T. Peacock, then mostly called basin tillage, has been revived recently (Musick, 1981). The practice involves placing dams or basins in the furrows. It was extensively evaluated in Kansas, Nebraska, and Oklahoma as a water conservation measure during the 1950s and early 1960s (Duley; Kuska and Mathews; Leubs; Daniel, Cox and Elwell; Locke and Mathews). New interest developed with promising results in 1975 using furrow dams with grain sorghum at Bushland, Texas, (Clark and Hudspeth, Clark and Jones) and with cotton at Lubbock, Texas (Bilbro and Hudspeth). Lyle and Dixon have also initiated research efforts using furrow dams at Halfway, Texas.

To avoid most runoff situations and integrate irrigation with rainfall occurrences, Stewart, Dusek, and Musick at Bushland, Texas, developed an irrigation practice with the acronym of LID (limited irrigation-dryland system). By using furrow dams, only the upper half of the field is fully irrigated. The next one-fourth of the field is the "tailwater runoff" section while the lower one-fourth is managed as the dryland portion of the field. However, the lower end is capable of retaining excessive runoff occurring from heavy rain storms or an unexpected advance of irrigation water. Substantial improvements in water and energy use efficiency have resulted from this approach. The LID, like many other methods of raising water and energy use efficiency, may result in decreased production per unit of land but will gain greater levels of production from the scarce energy and water resources.

Furrow dams, alternate row irrigation, and varied planting rates down the field are important facets of the LID system. All furrows are dammed but the furrow dams or basins in the alternate rows to be irrigated are cupped to allow water to accumulate and then breakover to the next furrow dam. Water is applied for a predetermined time to irrigate the upper half of the field adequately. The extent of the remainder of the field actually irrigated during the season depends on soil moisture and rainfall received during the irrigation period. Economic benefits of the system are three-fold: (1) increased efficiency of water and energy use through the conjunctive use of rainfall and irrigation, (2) near-automation and labor savings associated with irrigating, and (3) elimination of added costs of pumping tailwater or recirculating irrigation and rainfall runoff.

An analysis by Harman (1982a) indicated profits could be enhanced by $10 to $25 per additional acre of sorghum irrigated with the LID system. Water savings ranged from five to seven inches per acre (25 to 33 percent) when compared to the conventional program of fully irrigating less acreage of sorghum, recirculating 20 percent as tailwater, and devoting the remaining acreage on the farm to irrigated wheat. Energy requirements for the farm would be reduced by 25 to 33 percent based on typical irrigated wheat application rates of 20 to 24 inches of water in the Texas High Plains area.

The LEPA (Low Energy Precision Application) system, a mobile low pressure sprinkler system, also utilizes furrow dams with low pressure (1 to 4 psi) spray nozzles dropped near the soil surface (Lyle and Bordovsky). Furrow dams are used in conjunction with the system to prevent runoff. Energy requirements using furrow diking and the LEPA system are reduced by 12 to 25 percent when compared to conventional tillage with sprinkler systems of 20 and 50 psi respectively. Energy requirements per unit of water pumped were slightly higher for LEPA compared to furrow irrigation due to the increased water pressure needs. However, the application efficiency of the LEPA system approached 99 percent, resulting in lower irrigation costs per acre when compared to a furrow irrigation application efficiency of 82 percent.

Laser Leveling. A relatively new land forming practice, laser leveling, is being practiced in conjunction with irrigation in the arid Southwest at the current time. Hinz and Halderman describe the practice as leveling the land by means of a laser beam which is transmitted from a central command point to a receiver mounted on a scraper blade. The signal maintains the desired grade on the field by automatic control of hydraulic cylinders. Thus, furrow irrigated fields can be planed to a desired slope or border irrigated fields can be planed to a dead level situation. Water application efficiencies are estimated to improve by over 50 percent (Parsons). Yield impacts have not been documented, but observations indicate some yield increases are occurring. Daubert and Ayer concluded that the high cost of laser leveling, from $100 to $600 per acre depending on the situation plus $10 per acre per year for leveling maintenance, would pay in most cases by reducing the costs of irrigation. Cost sharing payments by the ASCS and the ability to reduce federal income taxes are important considerations in determining the amount of land to laser level each year.

Minimum Tillage and Innovative Crop Rotations

Some intriguing and innovative research has been accomplished by agronomists, soil scientists, weed scientists, and agricultural engineers to combat rapidly escalating tillage fuel costs in recent years. While most of the research has focused on minimizing energy requirements of tillage operations, an important side-effect of the research has been soil moisture conservation and control of soil erosion attained while reducing tillage costs.

Judicious use of herbicides with crop rotations appears to offer some promise in reducing irrigation water and energy requirements. Through proper weed control, reduced irrigation requirements and complementary crop-to-crop use of residual soil moisture and plant nutrients are possible. These advantages relate to all of western agriculture where water use requirements of crops are largely met with irrigation.

Musick, Wiese, and Allen developed an irrigated wheat/irrigated no-till sorghum/fallow rotation. The results of four years of research are summarized in Table 6.2. Soil moisture stored by maintaining wheat stubble with no-till during the 11-month idle season exceeded disked fields by 1 to 4 inches each year during the four year period. Sorghum yield improvement with no-till versus computational practice ranged from a low of 313 pounds/acre in 1970 with 12 inches of irrigation water to a high of 1,133 pounds/acre with two seasonal irrigations totaling less than 9 inches of water in 1973. The energy use

Table 6.2
Effects of disk and no-till methods on irrigated sorghum yields in a wheat/sorghum/fallow rotation, Bushland, Texas

		Irrigation Water			
Year[a]	Tillage Method	Preirrigation (inches)	Seasonal (inches)	Sorghum Yield (lbs/ac)	Energy Use Efficiency[b] (lbs/mcf)
1970	Disk		6	3,381	867
	No-Till		6	3,818	979
	Disk		12	5,200	667
	No-Till		12	5,513	707
1971	Disk	3	6	3,898	666
	No-Till	3	6	5,272	901
	Disk	3	12	5,441	558
	No-Till	3	12	6,012	617
1973	Disk	2.7	7.0	4,264	676
	No-Till	0	8.6	5,397	965
1974	Disk	4.9	0	3,336	1,047
	No-Till	0	4.7	4,264	1,396

[a]Level borders were utilized in 1970 and 1971 whereas graded furrows were utilized in 1973 and 1974.
[b]Based upon .65 mcf of natural gas used per inch of water.

efficiency under no-till increased over conventional practice in those same years by 40 lbs/mcf and 289 lbs/mcf, respectively. The percentage of energy use reduction per pound of sorghum produced by no-till practices ranged from a low of 6 percent to a high of 25 percent, including the 1974 production year.

The irrigation energy saved is not the only energy savings attributable to the no-till practice. Allen, Musick, and Dusek evaluated tillage energy savings of a wheat/sorghum rotation. Nearly 7 gallons per acre of diesel were saved on the farm by the no-till practice versus conventional tillage.

Shipley and Osbourne evaluated the economic impacts of no-till irrigated sorghum following wheat. The analysis indicated higher preharvest costs in 1973 for chemical fallow (no-till). However, increased yields of 600 to 1,400 pounds for respective 12 and 6 inch irrigation applications, improved profits by $18 per acre and $5 per acre, respectively. Shipley and Osbourne did not consider additional costs of no-till equipment or machinery and equipment depreciation costs.

Comparing this no-till practice and crop rotation at 1984 costs and prices in Table 6.3 and using an average sorghum yield increase of only 750 pounds per acre, no-till irrigated sorghum saved $4.57 per acre preharvest costs but required $4.50 per acre additional harvesting expense. Irrigation and machinery depreciation savings of nearly $30 per acre added to $32 per acre additional income at the 1984 wheat price results in a long-run profit of about $62 per acre (Harman, 1982b). It should be noted that machinery depreciation savings can be significant in no-till sorghum since smaller tractors and/or spray rigs are substituted for relatively large and expensive conventional tillage equipment. In this rotation of no-till sorghum with wheat, additional investment costs for coulters mounted on a conventional planter amount to only $200 per row.

MISCELLANEOUS MANAGEMENT OPTIONS

Other promising areas which can increase water and energy use efficiency include: (1) development of drouth tolerant varieties, (2) adoption of lower water-using crops, and (3) plant or soil profile modification to reduce evaporation.

Table 6.3
Relative profitability of limited tillage in irrigated wheat/irrigated sorghum/fallow rotation, Texas High Plains

Item	Tillage System		Change
	Conventional	No-Till	No-Till (-) Conventional
Income:			
Yield, lbs/acre	6,500	7,250	750
Price, dollars/cwt	$ 4.32	$ 4.32	
TOTAL INCOME	$280.80	$313.20	$32.40
Expenses:			
Seed	$ 3.60	$ 3.60	
Fertilizer	18.75	22.50	
Herbicides	9.30	24.12	
Insecticides	5.00	5.00	
Custom Application	3.00	6.00	
Machinery FOLR[a]	19.78	6.58	
Irrigation FOLR[b]	83.04	74.39	
Tractor Labor	6.40	2.36	
Irrigation Labor[b]	11.25	10.00	
Interest on Operating Capital	3.58	4.58	
TOTAL PREHARVEST COSTS	$163.70	$159.13	$(4.57)
Harvest and Haul, $.60/cwt	$ 39.00	$ 43.50	
Depreciation, Machinery	35.42	10.83	
Depreciation, Irrigation[b]	50.88	45.58	
TOTAL COST	$289.00	$259.04	$(29.96)
Returns to Land and Management:	$ (8.20)	$ 54.16	$62.36

[a]FOLR indicates fuel, oil, lubrication, and repairs.
[b]Irrigation costs are based on 24 inches per acre for conventional till sorghum and 21.5 inches per acre for no-till sorghum.

Drouth Tolerant Varieties. Recent genetic modification of corn, cotton, and wheat varieties has increased yields relative to water requirements. Quisenberry is selecting among exotic cotton germplasm that demonstrate a resistance to wilting. The resistance to wilt is probably due to a superior root system. Hurd has developed spring wheats that exhibit superior root systems and increased drouth tolerance. Porter has released several improved varieties of winter wheat over the past decade. The most recent release averaged 15 percent increase in energy use efficiency over the previous variety during 1980, 1981, and 1982. This increase was realized over several irrigation levels ranging from only preirrigation to pre plus four where the four seasonal irrigations were applied in early spring and at jointing, heading, and grain fill stages of growth (Table 6.4). These are but a few examples of public and private endeavors by scientists to reduce water and energy requirements of crops.

Lower Water Using Crops. Shifting to lower water using crops is also another option to producers in the West. Alternatives such as sunflower, jojoba, guayule, guar, safflower, and barley are being evaluated (Lacewell

Table 6.4
Comparative yields and energy use efficiency of TAM-105 and Scout winter wheat

	Irrigation		Yield		
Year	Treatment	Application (inches)	Scout (bu/ac)	TAM-105 (bu/ac)	Energy Use Eff, TAM-105/Scout[a] (percent)
1980-82	Preirrigation (P)	5.25	40.4	45.7	13.1
1980	P + Joint (J)	12.60	47.7	53.7	12.6
1981	P + Boot (B)	10.00	58.4	66.9	14.6
1980-81	P + Grain Fill (GF)	10.00	54.1	60.9	12.6
1982	P + Early Spring (ES) + J	14.00	51.7	59.5	15.1
1980	P + J + GF	16.60	58.2	69.8	19.9
1981	P + B + GF	15.00	66.2	81.5	23.1
1982	P + Heading (H) + GF	13.00	62.1	62.9	1.3
1982	P + ES + J + H + GF	21.00	78.2	97.1	24.2

[a]Based on .65 mcf of natural gas/inch of irrigation water.

and Collins; Moore, Lacewell, and Griffin; Harman, 1982b; Harman, 1982c; Shipley and Regier; Unger; Bauder and Ennen). Moore, Lacewell, and Griffin indicated, for example, sorghum would need to yield about 4,200 pounds/acre to provide equal economic returns with 1,700 pounds/acre sunflower yields in the northern Texas Panhandle. Harman, et al. developed a sunflower production function for the same area. Relating irrigation water to yield of sunflower, they found that yields of over 1,900 pounds/acre could be expected with the same amount of irrigation water as 4,200-pound sorghum. Thus, using 1981 price relationships, sunflower was more profitable than sorghum at low levels of irrigation (10 inches or less). Sunflower profits were below sorghum profits at the higher irrigation levels of 13.5 and 17.5 inches per acre.

Profits per unit of irrigation energy at three irrigation levels are summarized in Table 6.5 for sorghum and

Table 6.5
Comparative profits per unit irrigation energy use by sorghum and sunflower

	Returns over Variable Costs[a]		Profit per Unit Energy[b]	
Irrigation Level (inches)	Sorghum ($/ac)	Sunflower ($/ac)	Sorghum ($/mcf)	Sunflower ($/mcf)
(P + 1) 9.5	$ 59.88	$75.27	$ 9.70	$12.19
(P + 2) 13.5	97.27	88.29	11.08	10.06
(P + 3) 17.5	115.11	90.83	10.12	7.99

Source: Moore, et al., and Texas Agricultural Extension Service.

[a]Based on .10/lb sunflower price and $4.70/cwt. sorghum price.
[b]Based on .65 mcf of natural gas per inch of water.

and had to be replaced often to prevent CO_2 fixation and reduced production.

sunflower. Fixed machinery and land costs are excluded from the profit analysis since similar expenditures for machinery and land ownership are required for each crop. Profit per unit of energy at the low level of irrigation is $12.19/mcf of natural gas for sunflower compared with $9.70/mcf for sorghum. At higher levels of irrigation, however, sorghum profits per unit of energy use exceed those for sunflower.

Plant and Soil Modification. Plant modification through the use of antitranspirants applied to the foliage has been evaluated to determine the impacts on plant water use. Wendt, et al. found that significant potential exists if chemicals or other means can be found that will close the leaf stomata or coat leaves to slow evapotranspiration. The materials, however, evaluated were generally short-lived

Soil modification through natural or chemical mulches, gravel layers, and plastic sheeting have also been evaluated. Deep plowing and chiselling are the most commonly practiced means of soil modification. Plowing results in inverting soil layers; it has been used where a dense, less permeable soil needs to be broken up to increase water storage (Burnett; Siddoway, Unger and Schneider). Chiselling causes a channel through which water can penetrate slowly permeable soils. Choriki found that 2 inch layers of pea gravel on the soil surface increased soil water storage by 60 percent. Wendt evaluated chemicals that stabilize clod mulches or form films on the soil surface to reduce soil water evaporation. He found that costs and logistic problems prevent these from being feasible at present.

Gains in energy use efficiency through plant and soil modification can be realized. However, the economic feasibility of such methods would be limited in western agriculture and applicable only to high value crops or unique small scale truck farming of specialty crops. Commercial and large scale crop farms would find these practices cost-prohibitive at present.

IMPLICATIONS FOR GENERAL ADOPTION

Innovations, inventions, alternative irrigation systems, and changes in cropping strategies are all subject to economic, environmental, and institutional factors when considered for widespread adoption. An additional important factor influential in affecting the rate of adoption is that of management required for the practice.

Economic parameters include the start-up or investment cost, the operating cost, and the income generating ability or eventual payoff. A technological advancement having a low investment cost with few environment, institutional, or management problems will usually be adopted at a rapid pace if it is profitable. For example, the low cost and simplicity of furrow dikers plus the obvious rainfall retention benefits led to quick acceptance by producers in the Texas High Plains. Minimum tillage practices are increasing with various management systems being applied in different regions of the U.S. (Stewart and Harman).

Institutional factors, however, such as acreage controls, production quotas, market limitations, and cost-sharing programs can inhibit or increase rates of adoption. For example, laser leveling, irrigation distribution systems, bench terracing, and other soil and water conservation measures have been historically cost-shared by the USDA to encourage adoption rates. On the other hand, minimum-tilled crop rotations including both a crop with price supports and acreage control and one without may not be readily adopted due to the lack of price support for the non-program crop in the rotation.

The management practices discussed in this chapter to increase energy use efficiency are generally applicable to western agricultural needs. Adoption rates have varied, though, either due to lack of knowledge, resistance to change, or institutional factors imposed on producers. Most are economically feasible and environmentally safe for the producer's future consideration.

SUMMARY

Several options for improving on-farm energy efficiency have been discussed. The wide range of these options--from improved timing and rates of irrigation applications to planting water and energy use efficient crops--is indicative of the scope of management alternatives available to producers in the West.

Water being a scarce resource throughout the region forces agriculture to efficiently allocate and use the resource. The options discussed herein generally produce higher returns per unit of energy expended for irrigation at the farm level than conventional practices. Soil moisture retention through various means coupled with more efficient irrigation practices and more energy efficient crops can reduce energy requirements for commercial agricultural production.

Optimum timing of irrigations is an effective way to raise energy use efficiency. For example, one seasonal irrigation properly timed on sorghum can result in 25 percent to over a 500 percent increase in energy use efficiency. Two irrigations when properly timed can result in 25 percent to over a 400 percent increase in energy use efficiency. Non-peak load scheduling of irrigations has been found to offer 10 to 20 percent reductions in energy costs. Practices such as irrigating alternate furrows and reducing length of run, an increase energy use efficiency by 20 percent, especially on coarse textured soils. Furrow diking in combination with improved irrigation practices or systems raises efficiency by a like amount. Soil moisture conservation through minimum tillage practices can save 10 to 40 percent of the energy required for irrigation. Adoption of higher yielding varieties and switching to lower water using crops also improve irrigation energy efficiency and may raise farmer profits.

Profitability of new and improved means of increasing energy use efficiency should remain a major consideration. Some methods such as plant modification through antitranspirants have not been found to be economically feasible at present costs and product prices. However, researchers in the scientific community continue to reach out for innovative ways which may, in the future, become feasible alternatives for western irrigated agriculture.

REFERENCES

Allen, R.R., J.T. Musick, and D.A. Dusek. "Limited Tillage and Energy Use with Furrow Irrigated Grain Sorghum." Transactions of the ASAE, 23: 346-350, 1980.

Anderson, R.L., D. Yaron, and R. Young. "Models Designed to Efficiently Allocate Irrigation Water Use Based on Crop Response to Soil Moisture Stress." TR-8, Colorado State Univ., Fort Collins, 1977.

Bauder, J.W. and M.J. Ennen. "Soil Water Use Efficiency of Sunflowers." North Dakota Farm Research, 39(1): 9-12, 1977.

Bilbro, J.D. and E.B. Hudspeth, Jr. "Furrow Diking to Prevent Runoff and Increase Yields of Cotton." Texas Agr. Exp. Sta. Prog. Rpt. 3436, 3 pp., 1977.

Box, J.E., W.H. Sletten, J.H. Kyle, and Alexander Pope. "Effects of Soil Moisture, Temperature, and Fertility on Yield and Quality of Irrigated Potatoes in the Southern Plains." Agr. J., 55: 494, 1963.

Buchleiter, G., D.F. Heerman, and H.R. Duke. "Electrical Load Management Alternatives for Irrigation." ASAE Paper No. 79-2556, New Orleans, LA, 1979.

Burnett, E. "Profile Modification for Improved Water Intake and Storage." Seminar-Modifying the Soil and Water Environment for Approaching the Agricultural Potential of the Great Plains, Kansas State Univ., Manhattan, March 17-19, 1969.

California Public Utilities Commission. "Summary Report to U.S. Department of Energy - Agricultural Load Management Program." San Francisco, California, March 31,

1982.
Choriki, R.T. "The Influence of Different Soil Types, Treatments, and Soil Properties - On the Efficiency of Water Storage." M.S. Thesis, Montana State Univ., Bozeman, 1959.
Clark, R.N. and E.B. Hudspeth. "Runoff Control for Summer Crop Production in the Southern Plains." ASAE Paper No. 76-2008, St. Joseph, MI, 1976.
Clark, R.N. and O.R. Jones. "Furrow Dams for Conserving Rainwater in a Semiarid Climate." Proc. ASAE Conf., Crop Prod. Conserv. in the 80's., pp. 198-206, St. Joseph, MI, 1980.
Criddle, Wayne D., Sterling Davis, Claude H. Pair, and Dell G. Shockley. "Methods for Evaluating Irrigation Systems." USDA Agr. Handbook No. 82, 1956.
Daniel, H.A., M.B. Cox, and H.M. Elwell. "Stubble Mulch and Other Cultural Practices for Moisture Conservation and Wheat Production at the Wheatland Conservation Experiment Station." Cherokee, OK, 1942-51, USDA-ARS Prod. Res. Rpt. No. 6., 44 pp., 1956.
Daubert, J. and H. Ayer. "Laser Levelling and Farm Profits." Tech. B-224, Univ. of Arizona and USDA-NRED, June 1982.
Duley, F.L. "Yields of Different Cropping Systems and Fertilizer Tests Under Stubble Mulching and Plowing in Eastern Nebraska." Neb. Agr. Exp. Sta. Res. Bull. 190., 53 pp., 1960.
Evans, R.G. "Improved Semiautomatic Gates for Cut-Back Surface Irrigation Systems." Transactions of the ASAE, 20(1): 105-108, 112, 1977.
Fischbach, P.E. "Design of an Automated Surface Irrigation System." Proc. ASCE Irrigation and Drainage Special Conference, Phoenix, Arizona, pp. 219-237, 1968.
Fischbach, P.E. and B.R. Somerholder. "Efficiencies of an Automated Surface Irrigation System With and Without a Runoff Re-Use System." Transactions of the ASAE 14(4): 466-470, 1971.
Fischbach, P.E. and H.R. Mulliner. "Every Other Furrow Irrigation of Corn." ASAE, St. Joseph, MI, 49085, 1972.
Garton, J.E. "Designing an Automatic Cut-Back Furrow Irrigation System." Okla. Agr. Exp. Sta. Bull. B-651, 20 pp., 1966.
Gilley, J.R., D.F. Heerman and L.E. Stetson. "Irrigation Management - Energy Irrigation Challenges of the '80's." Proc. ASAE 2nd National Irrigation Symposium,

pp. 127-140, Oct. 1980.

Grimes, D.W., V.T. Walhood, and W.L. Dickens. "Alternative--Furrow Irrigation for San Joaquin Valley Cotton." Calif. Agr., 22: 4-6, 1968.

Harman, W.L. "Economic Benefits of the LID System." Popular article prepared for release in Southwestern Farm Press, October 7, 1982a.

________. "Limited Tillage - Prospects and Potential for the Texas High Plains." North Plains Research Field Day Report, Etter, Texas, 1982b.

________. "A Potential 'Economic Window' for Barley Production in the Texas High Plains." Paper presented to North Plains Research Field Day, Etter, Texas, August 18, 1982c.

Harman, W.L., P.W. Unger, and O.R. Jones. "Sunflower Yield Response to Furrow Irrigation on Fine-Textured Soils in the Texas High Plains." MP-1521, Texas Agr. Exp. Station, College Station, Texas, Dec. 1982.

Harris, T.R., H.P. Mapp, and J.F. Stone. "Irrigation Scheduling in the Oklahoma Panhandle: An Application of Stochastic Efficiency and Optimal Control Analysis." T-160, Oklahoma State Univ., Stillwater, 1983.

Hart, W.E. and J. Borrelli. "Mechanized Surface Irrigation Systems for Rolling Lands." Univ. of California Water Resource Ctr. Contr., 133: 93, 1970.

Heerman, D.F., H.H. Shull, and R.H. Mickelson. "Center-Pivot Design Capacities in Eastern Colorado." Journal of the Irrigation and Drainage Division, ASCE 100(IR2): 127-141, 1974.

Heerman, D.F. and H.R. Duke. "Electrical Load Management and Water Management." Proceedings of the Irrigation Association Technical Conference, pp. 60-67, 1978.

Hiler, E.A. and R.N. Clark. "Stress Day Index to Characterize Effects of Water on Crop Yields." Transactions of the ASAE, 14(4): 757-761, 1971.

Hinz, Walter W. and Allan D. Halderman. "Laser Beam Land Leveling Costs and Benefits." Bulletin Q114, Coop. Extension Service, The Univ. of Arizona, Tucson, Nov. 1978.

Hobson, B. "Cultural Energy Requirements of Colorado Crops." Monograph, Fort Collins: Colorado State Univ., 1976.

Humphreys, A.S. "Automatic Furrow Irrigation Systems." Transactions of the ASAE 14(3): 466-470, 1971.

Hurd, E.A. "Phenotype and Drought Tolerance in Wheat." In Plant Modification for More Efficient Water Use, John

F. Stone (ed.), Elsevier Scientific Publishing Co., New York, 1975.

Jensen, M.E. and J.T. Musick. "Irrigating Grain Sorghums." Leaflet No. 511, ASDA, June 1962.

Kuska, J.B. and O.R. Mathews. "Dryland Crop-Rotation and Tillage Experiments at the Colby (Kansas) Branch Exp. Sta." USDA Cir. 979, 87 pp., 1956.

Lacewell, R.D. and G.S. Collins. "Implications and Management Alternatives for Western Irrigated Agriculture." Proceedings of Conference on Impacts of Limited Water for Agriculture in the Arid West, in preparation, 1982.

Larson, D.L. and D.D. Fangmeier. "Energy Requirements for Irrigated Crop Production." Proceedings of the International Conference on Energy Use Management, Tucson, Arizona, pp. 743-750, October 1977.

Leubs, R.E. "Investigations of Cropping Systems, Tillage Methods, and Cultural Practices for Dryland Farming at the Fort Hays (Kansas) Branch Exp. Sta." Kansas Agr. Exp. Sta. Bull. 449, 114 pp., 1962.

Locke, L.F. and O.R. Mathews. "Relation of Cultural Practices to Winter Wheat Production." USDA Cir. 917, Southern Great Plains Field Station, Woodward, Oklahoma, 54 pp., 1953.

Lyle, W.M. and J.P. Bordovsky. "Low Energy Precision Application (LEPA) Irrigation System." Transactions of the ASAE, 80-2069, 1981.

Lyle, W.M. and D.R. Dixon. "Basin Tillage for Rainfall Retention." Transactions of the ASAE, 20: 1013-1017, 1021, 1977.

Moore, D.S., R.D. Lacewell, and R.C. Griffin. "Comparative Economic Position of Sunflower Production in the High and Rolling Plains of Texas." MP-1508, Texas Agr. Exp. Station, July 1982.

Musick, J.T. "Irrigating Grain Sorghum with Limited Water." Proc. SCS Conservation Workshop, College Station, Texas, July 15-16, 1968.

________. "Precipitation Management Techniques--New and Old." Efficiency in the 80's, Proc. 8th Annual Conference of Groundwater Management Districts, Lubbock, Texas, Dec. 1981.

Musick, J.T. and W.L. Harman. "TAM-105 Wheat Yield Response to Alternative Irrigation Levels, Texas High Plains." Texas Agr. Exp. Station progress report in review, 1984.

Musick, J.T., A.F. Wiese, and R.R. Allen. "Management of Bed-Furrow Irrigated Soil with Limited- and No-Tillage Systems." Transactions of the ASAE, 20: 666-672, 1977.

Musick, J.T. and B.A. Stewart. "Reducing Irrigation Energy for Sorghum." Energy in Agriculture ... Use and Production, Crop Production and Utilization Symposium, Amarillo, Texas, Feb. 14, 1980.

Musick, J.T. and D.A. Dusek. "Alternate-Furrow Irrigation of Fine-Textured Soils." Transactions of the ASAE, 17: 289-294, 1974.

Musick, J.T., W.H. Sletten, and D.A. Dusek. "Evaluation of Graded Furrow Irrigation with Length of Run on a Clay Loam Soil." Transactions of the ASAE, 16: 1075-1080, 1084, 1973.

New, Leon. "Influence of Alternate Furrow Irrigation and Time of Application on Grain Sorghum Production." Texas Agr. Exp. Sta. Prog. Rpt. No. 2953, 1971.

Nicolaescu, I. and E.G. Kruse. "Automatic Cutback Furrow Irrigation System Design." Proceedings Amer. Society Civil Engr., J. of the Irrig. and Drain. Div., 97(IR3): 343-353, 1971.

Parsons, Walter. Soil Conservation Service, Tucson, Arizona, Personal Communication, January 1981.

Patrick, N.A. "Energy Use Patterns for Agricultural Production in New Mexico." Agriculture and Energy, New York: Academic Press, 1977.

Peacock, D.T. Patent Office Official Gazette. 408(1931): 543.

Porter, K.B. Personal Communication, 1982.

Quisenberry, J.E., et al. "Exotic Cottons as Genetic Sources for Drouth Resistance." Crop Sci., 21: 889-895, 1981.

Roberts, E.B. and R.M. Hagan. "Energy Requirements of Alternatives in Water Supply, Use, and Conservation, and Water Quality Control in California." Final report to Univ. of California, Lawrence Livermore laboratory under Purchase Order No. 9865335, Contract No. @-7405 Eng 48 with U.S. Energy Research and Development Administration, 1977.

Robins, J.S. and G.E. Domingo. "Some Effects of Severe Soil Moisture Stress Deficits at Specific Growth Stages in Corn." Agron. J., 45: 618, 1953.

________. "Moisture and Nitrogen Effects on Irrigated Spring Wheat." Agron. J., 54: 135, 1962.

Schneider, A.D., J.T. Musick, and D.A. Dusek. "Efficient Wheat Irrigation with Limited Water." Transactions of the ASAE, 12(1): 23-26, 1969.

Shipley, J.L. "Economic Considerations in the Irrigation of Grain Sorghum and Corn, Texas High Plains." Proc. 32nd Corn and Sorghum Research conference, Chicago, Illinois, pp. 143-156, Dec. 1977.

Shipley, J.L. and C. Regier. "Water Response in the Production of Irrigated Grain Sorghum, High Plains of Texas." MP-1202, Texas Ag. Exp. Station, College Station, Texas, 1975.

Shipley, J. and C. Regier. "Sunflower Performance-Dryland and Limited Irrigation." PR-3416, Texas Agr. Exp. Station, 1976.

Shipley, J.L. and J.E. Osbourne. "Costs, Inputs, and Returns in Arid and Semiarid Areas." Conservation Tillage, Proceedings of a National Conference, Soil Conserv. Soc. Am., Ankeny, Iowa, pp. 168-179, 1973.

Siddoway, F.H., P.W. Unger, and A.D. Schneider. "Soil Modification for Improving Plant-Water Relations and Effects on ET." Seminar on Evapotranspiration in the Great Plains, Bushland, Texas, March 23-25, 1970.

Stegman, E.C., J.T. Musick, and J.T. Stewart. "Irrigation Water Management." In Design and Operation of Farm Irrigation Systems, M.E. Jensen (ed.), ASAE, pp. 763-816, 1980.

Stetson, L.E., D.G. Watts, F.C. Corey, and I.D. Nelson. "Irrigation System Management for Reducing Peak Electrical Demands." Transactions of the ASAE 18(2): 303-306, 311, 1975.

Stewart, B.A., D.A. Dusek, and J.T. Musick. "A Management System for Conjunctive Use of Rainfall and Limited Irrigation of Graded Furrows." Soil Sci. Soc. Am. J. 45: 413-419, 1981.

Stewart, B.A. and W.L. Harman. "Environmental Impacts of Declining Water Supplies in Irrigated Agriculture." Proceedings of Conference on Impacts of Limited Water for Agriculture in the Arid West, in preparation, TA-17973, Texas Agr. Exp. Station, College Station, Texas, 1982.

Stewart, B.A. and W.L. Harman. "Environmental Impacts." Chapter 15 of Water Scarcity: Impacts on Western Agriculture, Ernest A. Engelbert with Ann Foley Schuering (eds.), Univ. of California Press, Berkeley, pp. 354-379, 1984.

Stewart, J.T., R.D. Misra, W.O. Pruitt, and R.M. Hagan. "Irrigating Corn and Grain Sorghum with a Deficient Water Supply." Transactions of the ASAE, 8(2): 270-280, 1975.

Stone, J.F., J.E. Garton, B.B. Webb, H.E. Reeves, and J. Keflemariam. "Irrigation Water Conservation Using Wide-spaced Furrows." Soil Science Society of America Journal, 43(2): 467-411, 1979.

Stringham, G.E. and J. Keller. "Surge Flow for Automatic Irrigation." Proc. Am. Soc. Civil. Irrig. and Drain. Div. Spec. Conf., Albuquerque, New Mexico. pp. 132-142, 1979.

Unger, P.W. "Planting Date Effects on Growth, Yield and Oil of Irrigated Sunflower." Agronomy Journal, 72: 914-916, 1980.

Wendt, C.W. "Crude Oils as an Evaporation Suppressant." Agronomy Abstracts, annual meetings, Am. Soc. of Agron., Detroit, Michigan, 1969.

Wendt, C.W., A.B. Onken, H.J. Haus, and W.O. Willis. "Soil Water Evaporation." Proc. of Great Plains ET Seminar, Bushland, Texas, March 23-25, 1970.

Whittlesey, Norman K. "Demand for Electricity by Pacific Northwest Irrigated Agriculture." Report to Bonneville Power Administration, January 1982.

Wilson, D.L. and H.W. Ayer. "The Cost of Water in Western Agriculture." ERS Staff Report No. AGES820706, NRED-USDA, Washington, D.C., July 1982.

Young, K.B. "Temporal Economic Analysis of Alternative Groundwater Conservation Methods. Southern High Plains of Texas." Groundwater Management in the Great Plains, Great Plains Agricultural Council Pub. 72, Water Resources Committee, pp. 37-49, 1974.

7 Energy Inputs on Western Groundwater Irrigated Areas

Ronald D. Lacewell
Glenn S. Collins

INTRODUCTION

Energy cost is a significant factor influencing farmer profits and, thus, acres irrigated, quantity of water applied, and cropping patterns. The influence of energy on irrigation is particularly important where groundwater must be pumped from deep beneath the land and/or high pressure irrigation water distribution systems are used. Since about 85 percent of the irrigated acres of the U.S. are in the 17 western states (CAST), the impact of energy adjustments on irrigation is an especially serious issue for this region. Much of the emphasis of this chapter will be on irrigation in the West, particularly the incidence of irrigation from groundwater.

MAGNITUDE AND VALUE OF IRRIGATION

Magnitude

Frederick and Hanson found that of the approximate 50 million acres irrigated in the West, 46 million are in cropland and 4 million in pasture. They reviewed the varied published statistics on irrigated agriculture and provide useful approximations that are derived from many sources. In addition, Sloggett (1979, 1982) has conducted detailed surveys for irrigation across the U.S. The data presented in this section are derived from several sources and different periods. This constitutes a major limitation that must be considered in any interpretation. Thus, the statistics should be viewed as indicating general direction rather than an absolute.

About 61 percent of the water for western irrigated agriculture comes from surface sources and 39 percent is groundwater (Table 7.1). Of the 50 million acres irrigated, 34 percent are in the Mountain States, 26 percent in Pacific, 22 percent in the Northern Plains and 18 percent in the Southern Plains. By region, the Northern and Southern Plains are predominantly irrigated with groundwater while the Mountain and Pacific State acres are irrigated mostly with surface water. Using different data sources for acres irrigated (Frederick and Hanson; Sloggett, 1982) and quantity of water applied (Frederick and Hanson), it can generally be concluded that water applied per acre is about 60 percent greater from surface sources than from groundwater. This difference is primarily due to the location of the land being irrigated, however, rather than to the source of water.

Acreage in the West irrigated from groundwater in 1980 was estimated at 25.6 million. This was about the same as 1977 and up from 22.3 million in 1974 (Sloggett, 1982). Estimated groundwater pumpage in 1975 was 56 million acre feet (Frederick and Hanson). Combining these data sources, about 2.2 feet of water are pumped for each acre irrigated. Table 7.1 shows irrigation water applied by source within the 17 western states. The Northern and Southern Plains irrigate 67 percent of the acreage from groundwater (Sloggett, 1982) but use only 40 percent of the water (Frederick and Hanson). Groundwater use per acre is about 1.3 feet in the Great Plains compared to 4 feet in the Mountain and Pacific Regions.

An indication of the growth in irrigation from groundwater sources is reflected in annual pumpage rates. For the 17 western states, annual groundwater pumpage rates have increased from 18.2 million acre feet in 1950 to the 56 million acre feet in 1975. The Northern Plains increased from .8 to 11.2 million acre feet, a 1,300 percent increase. The Southern Plains increased from 1.9 to 11.1 million acre feet, a 484 percent. This is compared to a 174 percent increase in the Mountain Region and an 87 percent increase in the Pacific Region (Frederick and Hanson).

The characteristics of the aquifers containing groundwater are highly diverse. They vary in depth to the static water level, depth to the base of the aquifer, relation to surrounding formations, and specific yield. Throughout the

Table 7.1
Acres irrigated and water applied from surface and groundwater sources, by region, in the 17 western states

Regions	Total Irrigated Acres	Total Water Applied	Surface Water Applied	Ground-Water Applied
	mil.	---- mil. acre foot -----		
Northern Plains	10.8	14.3	3.1	11.2
Southern Plains	9.0	13.9	2.8	11.1
Mountain	17.1	65.3	50.9	14.5
Pacific	13.3	50.9	31.6	19.3
TOTAL	50.2	144.5	88.4	56.1

Source: Frederick and Hanson.

West, mining of groundwater (extracting more than is recharged) exceeds 22 million acre feet per year (Sloggett, 1979). Groundwater mining, alternative uses, rising energy costs, and impairment of groundwater quality pose serious threats to the long term outlook for irrigation from groundwater in the West (National Water Commission).

About 60 percent of the acres served by groundwater are irrigated by gravity flow (flood or furrow irrigation), 20 percent with center pivot and 20 percent with other sprinklers such as side-roll (Sloggett, 1982). The trend is toward more sprinkler irrigation in many areas to achieve higher water use efficiency levels, but with the effect of increasing energy use.

Value

In 1978, 24 percent of the value of marketings from all cropland and rangeland in the U.S. came from irrigated land. Irrigated cropland was only 14 percent of total acreage harvested in 1978 yet produced $26 billion of products (U.S. Department of Commerce). Of the total value of crops and forest products sold in 1974, the proportion from irrigated lands was more than 90 percent in Arizona, California, Nevada, and New Mexico; 80 to 90 percent in Idaho, Utah, and Wyoming; and 60 to 80 percent in Colorado, Nebraska, Oregon, Texas, and Washington (CAST).

The benefits of irrigation can be defined as the increase in profit to the producer using irrigation compared to producer profit without irrigation. A study by Frank and Beattie updated by Young to 1982 dollars indicates that the value of irrigation at the margin is $10-$15 per acre foot in the Intermountain Valley (Upper Colorado River and Snake River Basin), $20-$25 in the desert Southwest and central California, and $40-$45 per acre foot in the Ogallala groundwater region of the High Plains (OTA, p. 389). According to Young, estimates reported by Howitt, et al. and Gollehon, et al. are very similar to those of Frank and Beattie. Lastly, a Department of Commerce study of water value in the Ogallala Region showed a value of $60 and $80 per acre foot for water used in irrigation.

IMPACT OF ENERGY PRICE

Energy Price and Pumping Cost

The primary fuels used for irrigation energy in the U.S. are electricity, diesel, gasoline, natural gas, and LPG. Table 7.2 indicates the percent of acres irrigated using these different fuels in the 17 western states. Electricity is most preferred, being used on 47 percent of the acres, an increase of 29 percent since 1974 (Sloggett). Natural gas is used on 26 percent of the acres, increasing by 6 percent since 1974. Although diesel is a far third at 18 percent, acreage irrigated using diesel fuel has increased 97 percent since 1974. Acres irrigated using gasoline and LPG have declined by 33 percent and 24 percent respectively since 1974. Both can be considered as minor fuels.

Thus, price adjustments in electricity, natural gas, and diesel can be expected to have significant impacts on irrigation. Table 7.3 shows acres irrigated by fuel type and region in the western U.S. The Northern Plains is a relatively large diesel user for irrigation but also relies heavily on electricity and natural gas. The Southern Plains predominately uses natural gas while the Mountain and Pacific States emphasize electricity. Thus, the impact

Table 7.2
Percent of acreage irrigated with alternative fuels in the 17 western states, 1980

Fuel	Acres Irrigated (percent)
Electricity	47.4
Diesel	18.1
Gasoline	2.3
Natural Gas	26.3
LPG	5.9

Source: Sloggett (1982).

Table 7.3
Acres irrigated with alternative fuels by region, 1980[a]

Fuel	Northern Plains	Southern Plains	Mountain	Pacific
	million acres			
Electricity	3.3	2.1	4.5	6.7
Diesel	2.8	0.2	0.3	0.1
Gasoline	0.1	0.1	0.1	0.0
Natural Gas	3.6	6.2	1.1	0.1
LPG	1.3	0.5	0.2	0.0

Source: Sloggett, 1982.

[a]Values are for on-farm pumped water only.

of a price change in a specific fuel will not have the same effect on irrigation in the different regions.

Natural Gas. About 35 percent of the acres irrigated in the West, mainly in the Southern Plains, use natural gas for pumping (Sloggett, 1982). The quantity of natural gas in thousand cubic feet, required to pump one acre inch is given in the following equation.

$$NG = [(2.31 * PSI) + LIFT] * [0.11427/(PE * DE)] * (.002544/EE)$$

where:

NG = thousand cubic feet (mcf) of natural gas required to pump one acre inch of water
PSI = irrigation system operation pressure in pounds per square inch
LIFT = depth in feet to the static water level from the surface
PE = pump efficiency (decimal)
DE = drive efficiency (decimal)
EE = engine efficiency (decimal)

This equation was applied for alternative scenarios to estimate the effect on fuel consumption and cost per acre foot of water pumped using natural gas as the fuel, see Table 7.4. In this case, a drive efficiency of 97 percent and engine efficiency just above 23 percent were used. The effect of operating pressure suggests going from 90 PSI to 10 PSI for operating the irrigation system, with 250 feet of lift and a 65 percent pump efficiency would reduce natural gas use 41 percent. This reduction in sprinkler operating pressure results in an $18 per acre foot decline in pumping costs. Also significant is the effect of pump efficiency. By increasing pump efficiency from 50 percent to 70 percent given 45 PSI and 250 feet of lift reduces energy use 28 percent or $12.40 per acre foot.

Diesel. About 11 percent of the acres irrigated in the West use diesel fuel for pumping (Sloggett, 1982). An equation for estimating diesel fuel use per acre inch of water as given by Kletke, Harris, and Mapp is as follows:

$$D = [(2.31 * PSI) + LIFT] * [.008683/PE * DE)]$$

where:

Table 7.4
Fuel required and cost per acre foot of water pumped using natural gas

		Assumptions		
Natural Gas (mcf)	Cost[a] ($ ac.ft.)	PSI (pounds)	LIFT (feet)	Pump Efficiency (percent)
6.5	26.00	10	250	65
8.5	34.00	45	250	65
11.0	44.00	90	250	65
11.0	44.00	45	250	50
7.9	31.60	45	250	70
1.7	6.80	10	50	65
3.7	14.80	45	50	65
12.5	50.00	10	500	65

[a]Based on $4.00 per mcf.

D = diesel use in gallons per acre inch of water pumped

Table 7.5 shows expected diesel use and associated costs under the same scenarios as previously considered for natural gas. With diesel prices at $1.00 per gallon and natural gas at $4.00 per mcf, natural gas appears to be much less costly. Generally, the cost of pumping with natural gas is around 40 percent less than with diesel. The effects of pressure, lift, and pump efficiency on pumping cost are similar regardless of fuel type. This suggests some opportunities for reducing pumping costs by converting to low pressure systems and repairing pumps to increase efficiency.

Electricity. Approximately 50 percent of the acres irrigated in the West using energy for pumping depend upon electricity for power (Sloggett, 1982). With electricity, engine efficiency is usually very high (above 90 percent)

Table 7.5
Fuel requirements and cost per acre foot of water pumped using diesel

		Assumptions		
Diesel (gal)	Cost[a] ($/ac.ft.)	PSI (pounds)	LIFT (feet)	Pump Efficiency (percent)
43.24	43.24	10	250	65
56.04	56.04	45	250	65
72.50	72.50	90	250	65
72.85	72.85	45	250	50
52.04	52.04	45	250	70
11.57	11.57	10	50	65
24.37	24.37	45	50	65
82.82	82.82	10	500	65

[a]Based on $1.00 per gallon.

and drive efficiency is 100 percent. Thus, an equation for estimating kilowatt hours (kwh) per acre inch of water pumped is as follows (Kletke, Harris, and Mapp):

$$kwh = [(2.37 * PSI) + LIFT] * 0.101141/PE)$$

where:

kwh = electricity in kilowatt hours per acre inch of water pumped.

Table 7.6 provides estimates of electricity use and cost per acre foot of water pumped. With electricity priced at $0.05 per kwh, the costs of pumping are approximately equal to natural gas priced at $4.00 per mcf. However, these estimates are applicable only to the lift, PSI and efficiencies assumed. The reader may easily calculate the effect of assuming energy costs different from those used in Tables 7.4, 7.5, and 7.6.

Table 7.6
Fuel requirements and cost per acre foot of water pumped using electricity

		Assumptions		
Electricity (kwh)	Cost[a] ($/ac.ft.)	PSI (pounds)	LIFT (feet)	Pump Efficiency (percent)
504	25.20	10	250	65
653	32.65	45	250	65
844	42.40	90	250	65
849	42.45	45	250	50
606	30.30	45	250	70
135	6.75	10	50	65
284	14.20	45	50	65
965	48.25	10	500	65

[a]Based on $0.05 per kwh.

To characterize the groundwater irrigation cost situation in the West, Table 7.7 indicates the typical range of pumping lifts by state. These data suggest generally higher costs of pumping in Arizona, California, Idaho, and Washington than in other states. However, in all states the aquifer depths are very heterogenous. Most states have some wells with lifts less than 75 feet and more than 400 feet of lift.

Adjustments in Irrigated Regions

Input prices, relative to product prices, were lower for farmers in the 1950s and 1960s than in recent years (Lacewell and McGrann). Before 1970, the maximum profit level was near the maximum yield level for most crops. This applied to both irrigated and non-irrigated crop production. However, with rapidly rising energy costs,

Table 7.7
Pump lift ranges for major groundwater irrigated states in the West

State	Pumping Lift (feet)
Arizona	75-535
California	75-300
Colorado	175-270
Idaho	50-400
Kansas	175-250
Nebraska	25-250
New Mexico	75-225
Oklahoma	200-275
Texas	75-225
Washington	75-500

Source: Sloggett, 1979, Table 4, p. 13.

fertilizer prices and chemical costs it is frequently undesirable or unprofitable to strive for maximum yields.

Typically, irrigated crop production is more energy intensive and cost per unit of output greater than for dryland production. This makes the irrigation farmer highly sensitive to input prices. As the cost to pump water increases relative to crop prices, there is an economic incentive to apply less water per acre. As an example, given a corn price of $1.20/bu as in the 1960s and a water cost $0.30/acre inch, the farmer would use 28.8 inches of water to maximize profit in Texas. With corn priced at $3.00/bu but pumping costs at $4.00/acre inch, the profit maximizing level of water use declines to 25.5 inches (Lacewell and McGrann).

There are selected examples where rising input costs have severely affected irrigated agriculture. The 1972-1975 natural gas price rise of 450 percent in the Trans Pecos of Texas contributed to irrigated cotton declining from over 200,000 acres to less than 20,000 acres (Clarke). This phenomenon implies that the effects of declining groundwater and increasing energy costs will be regional in nature. Those areas with the deepest water will be the first impacted by energy cost increases, as in the Trans Pecos example. However, adjustments due to declining groundwater generally are expected to be very gradual. The effect of rising energy prices depends on the magnitude of the price increase, pump lift (depth to water), distribution system, operating pressure, and crop opportunities. Thus emerges the issue of national implications.

A national analysis considered the effect increased water prices, as they might be affected by rising energy costs, on U.S. agriculture (Christensen, Morton, and Heady). When only groundwater prices were increased, the major reduction in irrigated acreage was in Missouri, Arkansas-White-Red, Texas-Gulf, and Colorado River Basins. Irrigated acreage declined 17 percent as a result of the groundwater price increases. When only surface water price was increased, irrigated acreages declined 18 percent. The reduction in irrigated acreage occurred primarily in the Missouri, Arkansas-White-Red, and California River Basins with little change in the other regions. The study further indicated that large increases in water prices are not expected to result in great geographical changes in irrigated cropland.

A more recent national analysis of the expected effects of rising costs on agriculture was done by Lacewell and Collins. The impacts on U.S. agriculture, both regional and national, of increasing water costs were examined using an econometric model (TECHSIM) developed at Texas A & M (Collins). TECHSIM is a regional field crop and national livestock model designed to evaluate yield and/or production cost changes. Within TECHSIM, extraneous information is used as a basis to reflect changes in supply and/or demand. This means data regarding the expected change in per acre yield and/or cost of production by region and crop associated with a specified policy or technology is incorporated into the model for further analysis. Thus, the expected impacts of a yield and/or cost change can be estimated.

In TECHSIM the U.S. is represented by 13 field crop producing regions shown in Figure 7.1 and four national livestock categories. The field crops included in the model are: corn, small grains (wheat barley, and oats), grain sorghum, cotton lint, cottonseed, and soybeans. The model also contains the forward meal and oil products of cotton seed and soybeans. The national livestock products

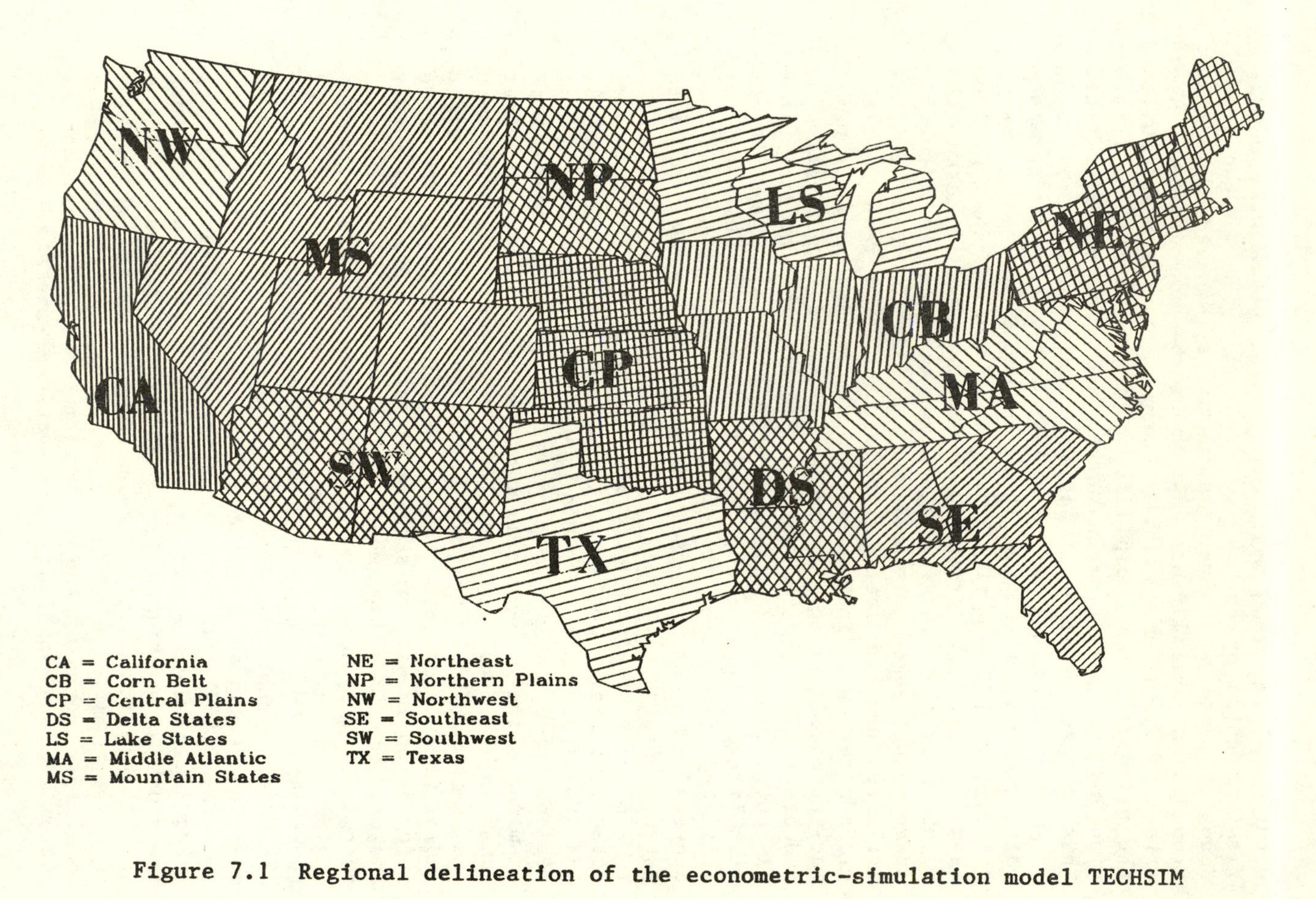

Figure 7.1 Regional delineation of the econometric-simulation model TECHSIM

included in the model are fed beef, non-fed beef, pork, and sheep.

TECHSIM is a recursive model which centers on three types of estimates: planted acreage, yield, and demand. The model assumes that producers within a region make planting decisions based on expected net returns of various field crops within the region. The recursiveness of TECHSIM is based upon the expected net returns which are equivalent to the previous year's net returns. The yield equations reflect productivity changes of the model's base period, 1961-1977.

The demand equations, with the exception of final consumer demands of the livestock sector, are intermediate demands. This implies that additional processing of these commodities is required before they reach the final consumer. Effects of increasing ground and surface water costs are introduced into the model by changing each crop's variable per acre production cost for each region. The model traces these impacts and provides estimates of regional and national planted acres, yields, production, and supplies of each field crop. The results also give estimates of national demands (domestic, export, private, and government stocks) of all commodities in the model.

The model is based on 1979 costs and returns. Average costs of groundwater were about $17 per acre foot and surface water about $7 in 1979 across the West (U.S. Department of Agriculture). Two national water alternatives are examined here with TECHSIM. The first is doubling the per acre costs of both ground and surface water (Scenario I). The second is doubling the cost of groundwater but letting surface water increase to $40 per acre foot (Scenario II). These two alternatives do not reflect any particular policy but rather provide insight into expected adjustments under these specific changes in water costs.

Planted Acreage Shifts. Regional changes in planted acres due to the two increased water cost scenarios are shown in Table 7.8. All results reflect the final impact of increasing ground and surface water costs to U.S. agriculture. Total U.S. planted acres decreased under both alternatives for all crops except small grains and grain sorghum. Grain sorghum acreage actually increased slightly in both cases. The largest percentage shifts occurred in the western regions where irrigation is predominant in agricultural production. The greatest regional percentage shift by crop was for corn in the Mountain States (MS)

Table 7.8
Regional shifts in planted acres and net returns[a]

Policy	Regions													
	US	NW	CA	MS	SW	CP	NP	TX	LS	CB	DS	SE	MA	NE
Double Ground and Surface Water costs														
Planted Acres:														
Corn (1000)	-44	1	1	-162	1	-419	10	-40	45	75	0	8	-7	7
Percent Change[b]	-.7	.2	.4	-27.2	1.2	-5.6	2.2	-4.3	.4	.2	0	.3	-.2	.3
Small Grains (1000)	4	-23	25	142	22	-44	0	1	-37	9	1	1	1	1
Percent Change[b]	.0	-.5	1.1	1.1	2.7	-.1	.0	.1	-.4	.1	.1	.1	.1	.1
Grain Sorghum (1000)	18	0	-32	0	86	100	0	17	0	7	3	-1	-2	0
Percent Change[b]	1.2	.0	-10.3	.0	22.1	1.5	.0	.3	.0	1.0	.9	-.9	-1.2	.0
Cotton (1000)	-4	0	-26	0	-89	15	0	32	0	0	14	10	3	0
Percent Change[b]	-.3	.0	-2.2	.0	-16.4	2.8	.0	.6	.0	.0	.4	1.1	.6	.0
Soybeans (1000)	-7	0	0	0	0	9	2	-1	50	-118	-38	6	16	4
Percent Change[b]	-.1	.0	.0	.0	.0	.3	.4	-.2	11.3	-4.0	-.4	.2	.3	.6

(Continued)

Net Returns (m. $)	-318	-24	-139	-129	-144	-233	13	-200	81	303	68	30	38	18
Percent Change[b]	-1.4	-18.0	-23.9	-29.9	-39.7	-7.7	2.4	-13.2	3.6	2.7	3.9	7.4	5.1	3.7
Double Ground Water & Increase Surface Water to $40/ac. ft.														
Planted Acres:														
Corn (1000)	-47	1	-4	-155	1	-389	10	-38	41	70	0	8	7	7
Percent Change[b]	-.7	1.8	-1.5	-26.0	1.2	-5.2	.5	-4.1	.4	.2	.0	.3	.2	.3
Small Grains (1000)	10	-31	0	101	13	-37	0	2	-33	16	2	1	2	2
Percent Change[b]	.1	-.6	.0	.8	1.6	-.1	.0	.0	-.4	.2	.3	.1	.1	.1
Grain Sorghum (1000)	18	0	-31	0	85	104	0	14	0	7	3	-1	-2	0
Percent Change[b]	1.2	.0	-9.9	.0	21.9	1.6	.0	.2	.0	1.0	.9	-.9	-1.2	.0
Cotton (1000)	-5	0	-38	0	-80	5	0	33	0	0	15	11	4	0
Percent Change[b]	-.4	.0	-3.2	.0	-14.8	.9	.0	.6	.0	.0	.4	1.2	.8	.0
Soybeans (1000)	-7	0	0	0	0	7	2	-2	49	-111	-42	5	15	3
Percent Change[b]	-.1	.0	.0	.0	.0	.3	.4	-.3	1.1	-.4	-.4	.1	.3	.5
Net Returns (m. $)	-547	-34	-397	-162	-160	-233	17	-211	81	297	70	30	38	17
Percent Change[b]	-2.4	-24.8	-51.1	-37.6	-44.2	-7.6	3.0	-13.9	3.6	2.6	4.0	7.4	5.0	3.7

[a] Regional delineations are illustrated in Figure 7.1.
[b] Percent change from baseline.

Region. In this region corn acreage decreased by 27.1 percent under Scenario I and decreased by 26 percent with Scenario II.

Net-Return Shifts. The total regional and national short-run net returns, farmer profits, in 1982 dollars are also shown in Table 7.8. For each region these net returns represent the sum of the changes in net returns for all field crops in a region. The national net return figures represent the summation of all regions' net returns.

Nationally, total net returns decrease for field crops. Regional net return decreases are observed for Northwest (NW), California (CA), Mountain States (MS), Southwest (SW), Central Plains (CP), and Texas (TX) Regions. The largest decrease in net returns for Scenario I occurs for the Southwest (SW) Region. However, with Scenario II, California (CA) becomes the largest looser on an absolute and percentage basis. The Corn Belt (CB) Region gains the largest dollar amount from increased water costs. The Corn Belt gain amounts to approximately $300 million for both water alternatives. However, from a percentage standpoint the Corn Belt Region ranks last of the seven regions which show positive increases in regional net returns. Nationally, total net returns decrease by 1.4 and 2.4 percent in the two water scenarios.

Price Shifts. The price shifts associated with increased water costs are modest. Price increases occur for corn, cotton lint, cottonseed, and soybeans. The largest price increase is for corn and soybeans. However, these prices increased by less than $.12 per bushel. Small grains and grain sorghum have slight price decreases. These small price changes are to be expected, given the small change in irrigated acreage of the field crops. The impact of increased water costs to the livestock sector are also small. Increases in livestock prices are observed for fed beef, pork and sheep, and lambs while only non-fed beef decrease in farm price. This is because feed prices of corn, soybean meal, and cottonseed meal increase proportionately more than small grains and grain sorghum decrease. Hence, livestock producers would shift to non-fed animal units which tend to increase fed animal prices and decrease non-fed animal prices.

Annual national net farm income would decrease over $.9 billion with Scenario I water prices and over $1.2 billion when Scenario II water costs are imposed. Very few agricultural industries gain as a result of these water alternatives. Consumers of field crops lose the most.

However, U.S. field crop producers and soybean and cottonseed meal and oil industries also lose. The livestock sector of the U.S. economy would gain; however, final consumers of livestock products lose.

The major implication of this analysis is that if water costs increase due to rising energy costs, western regions which are heavily dependent upon irrigated crop production will have the greatest decrease in net returns. Furthermore, additional losses will result in areas such as California if average surface water costs increase. Nationally, these water cost increases will result in only modest decreases in planted acres and small changes in prices of farm products. However, other industries besides field crop producers stand to lose with increasing water costs, so that the total annual loss would approach $1 billion under the second scenario.

This analysis was based on a uniform increase in energy and water costs across all regions. This is not a likely scenario since energy prices adjust differently among fuels and among regions.

OUTLOOK FOR GROUNDWATER IRRIGATION

Rising energy prices and declining groundwater supplies create uncertainty about the magnitude and stability of irrigation from groundwater. The previous section indicated the effect of lift and other factors on fuel required to pump an acre-foot of water.

Table 7.9 shows the acres in major western states that are irrigated from declining aquifers and their average annual rate of decline. About 15 million acres of western U.S. groundwater irrigated lands are incurring declining water levels in excess of one-half foot per year (Sloggett, 1982). This is primarily in the Southern Plains where water is drawn from the Ogallala aquifer. Mining from this aquifer is estimated at 14 million acre feet per year (Frederick and Hanson).

Higher energy prices and declining water supplies are expected to cause a reduction in irrigated acres in some regions. Those areas most vulnerable include Arizona, California, Idaho, Kansas, Texas, and the Oklahoma Panhandle. Predominant crops affected include cotton, citrus, grapes, grain sorghum, and rice (Sloggett, 1979). In any case, adjustments in groundwater use, due to rising

Table 7.9
Western U.S. acres irrigated from groundwater and percent of acres with declining groundwater supplies

State	Groundwater Irrigated	Declining Groundwater Acres	Percent	Average Annual Decline
	------- 1000 Acres		-------	--feet--
Arizona	940	734	78	2-3
California	4,388	1,814	41	2-5
Colorado	1,650	570	35	2
Idaho	1,149	150	12	2-5
Kansas	3,083	1,995	65	1-4
Nebraska	5,855	1,842	32	.5-2
Oklahoma	730	507	70	1-2.5
Texas	7,846	6,425	82	1-4
TOTAL	26,400	14,597	55	N/A

Source: Sloggett, 1979, Table 1, p. 9 and Table 4, p. 13.

energy prices and/or declining groundwater supplies, are expected to be gradual (Petty, et al.).

Of the groundwater mining in the Ogallala, some 63 percent is occurring in Texas and Oklahoma. However, groundwater mining does not threaten to totally exhaust the water stored in the western water resource regions in the foreseeable future (Frederick and Hanson). A major study completed in 1982 for the Ogallala over a six-state region projects that from 1977 to 2020 some 24 percent of the water in the aquifer will be depleted but irrigated acres will increase by 26 percent (High Plains Associates). For Texas, New Mexico, and Oklahoma, irrigated acres will decline by 17 percent and 66 percent of the aquifer will be depleted; while for Nebraska, Kansas, and Colorado, irrigated acres will increase by 66 percent and with only an 18 percent depletion of the aquifer. It is noted that groundwater mining has, particularly in parts of the Ogallala, depleted available groundwater so that only household water is now available. However, it is estimated there are sufficient quantities of groundwater available for irrigation beyond the year 2020, even in areas of rapid decline as the Texas High Plains. This means as some areas lose irrigated acres others will be increasing to offset the reduction. Food production capacity from irrigated agriculture shows no signs of decreasing in the near future.

A last major factor affecting irrigation is crop prices. Frederick and Hanson project that with no change in real crop prices, irrigated acres in the West will increase from 50.2 million to 53.5 million by 2010. With a 25 percent increase in real crop prices, the projected irrigated acreage is 58.0 million. Frederick and Hanson project a two million irrigated acre reduction in the Southern Plains, little change in the Mountain Region, a small increase in the Pacific Region, and major increases in the Northern Plains. Thus, the increases in irrigated acres are expected to be in areas that use groundwater. According to these projections, the long-term elasticity of irrigation with respect to crop prices is only about 0.3, or a 25 percent increase in crop prices increases irrigation only 8 percent. As stated by Frederick and Hanson, this lack of response of irrigated acres to crop price is explained partially by limited water supplies, vastly superior economic position of municipal and industrial users in bidding for water, and the importance of institutional factors affecting management and use.

REFERENCES

Cast. "Water Use in Agriculture: Now and for the Future." Council for Agricultural Science and Technology Report No. 95, September 1982.

Christensen, Douglas A., Andrew Morton, and Earl O. Heady. "The Potential Effect of Increased Water Prices on U.S. Agriculture." Center for Agricultural and Rural Development, CARD Report No. 101, Iowa State Univ., 1981.

Clarke, Neville P. "Texas Agriculture in the 80's: The Critical Decade." Texas Agricultural Experiment B-1341, Texas A&M Univ., 1980.

Collins, Glenn S. "An Econometric Simulation Model for Evaluating Aggregate Economic Impacts of Technological Change on Major U.S. Field Crops." Unpublished Ph.D. dissertation, Texas A&M Univ., 1980.

Department of Agriculture. "FEDS Budgets." Economics, Statistics and Cooperative Service, Department of Agricultural Economics, Oklahoma State Univ., Stillwater, 1981.

Frank, Michael D. and Bruce R. Beattie. "The Economic Value of Irrigation Water in the Western United States: An Application of Ridge Regression." Texas Department of Water Resources TR-99, Texas A&M Univ., 1979.

Frederick, Kenneth D. and James C. Hanson. "Water for Western Agriculture." Resources for the Future, Washington, D.C., 1982.

Gollehon, N.R., et al. "Impacts on Irrigated Agriculture for Energy Development in the Rocky Mountain Region." Southwest Review for Management and Economics, 1(1:61-88), 1981.

High Plains Associates. "Six-State High Plains Ogallala Aquifer Regional Resources Study." Final report to the U.S. Department of Commerce, 1982.

Howitt, R.E., D.E. Mann, and H.J. Vaux, Jr. "The Economics of Water Allocation." In Competition for California Water: Alternative Resolutions, Earnest A. Englebert, et al., Berkeley, University of California Press, 1982.

Kletke, D.D., T.R. Harris, and H.P. Mapp, Jr. "Irrigation Cost Program Users Reference Manual." Oklahoma State Univ. Research, Report P-770, Stillwater, 1978.

Lacewell, Ronald D. and Glenn S. Collins. "Implications and Management Alternatives for Western Irrigated Agriculture." Paper presented at Conference on Impacts of Limited Water for Irrigated Agriculture in the Arid West, Asilomar Conference Center, Pacific Grove, California, Sept. 28 - Oct. 1, 1982.

Lacewell, Ronald D. and James M. McGrann. "Research and Extension Issues in Production Economics." Southern Journal of Agricultural Economics, 14(1): 65-74, 1982.

National Water Commission. "New Directions in U.S. Water Policy: Summary, Conclusions and Recommendations." Final Report of National Water Commission, Superintendent of Documents, U.S. Government Printing Office, Washington, D.C., 1973.

Office of Technology Assessment. "Water Related Technologies for Sustaining Agricultural Lands in U.S. Arid/Semiarid Land." Washington, D.C., U.S. Congress, OTA-F-212, October 1983.

Petty, James A., Ronald D. Lacewell, Daniel C. Hardin and Robert E. Whitson. "Impact of Alternative Energy Prices, Tenure Arrangements and Irrigation Technologies on a Typical High Plains Farm." Texas Water Resources Institute TR-106, Texas A&M Univ., 1980.

Sloggett, Gordon. "Energy and U.S. Agriculture: Irrigation Pumping, 1974-77." Agr. Econ. Report No. 436, Washington, D.C., U.S. Dept. of Agriculture, September 1979.

Sloggett, Gordon. "Energy and U.S. Agriculture: Irrigation Pumping, 1974-80." Agr. Econ. Report No. 495, Washington, D.C., Dept. of Agricultural, December 1983.

U.S. Department of Commerce. "1978 Census of Agriculture." Vol. 1, Summary and State Data, U.S. Government Printing Office, Washington D.C., 1981.

U.S. Water Resources Council. "The Nations Water Resources." Vol. 1, U.S. Printing Office, Washington, D.C., 1975.

Williford, George, Bruce R. Beattie, and Ronald D. Lacewell. "The Impact of a Declining Groundwater Supply in the Northern High Plains of Texas and Oklahoma on Expenditures for Public Services." Texas Water Resources Institute TR-71, Texas A&M Univ., 1976.

Young, Robert A. "Direct and Regional Economic Impacts of Competition for Irrigation Water in the West." Paper presented at conference on Impacts of Limited Water for Irrigated Agriculture in the Arid West, Asilomar Conference Center, Pacific Grove, California, Sept. 28 - Oct. 1, 1982.

8
Energy Used for Pumping Irrigation Water

Gordon R. Sloggett

INTRODUCTION

Energy used for pumping irrigation water attracted little attention until the vulnerability of our energy supplies were exposed by the Arab oil embargo in 1973. Relatively low energy costs for pumping irrigation water contributed to the lack of interest in irrigation energy problems. Comprehensive statistics on energy used for pumping irrigation water were not compiled until 1974. The oil embargo and subsequent energy price increases sparked an interest in knowing the type, quantity, location, and cost of irrigation pumping energy.

Irrigation water generally derives from two sources, underground aquifers and surface water found in lakes, reservoirs, rivers, and ponds. Groundwater sources include some artesian wells that flow without pumping, but most groundwater used for irrigation requires pumping. Surface water sources include man-made reservoirs or stream diversion dams where irrigation water is delivered to farms by gravity with no pumping required, but many surface water sources do require some pumping for delivery to farms and ranches.

Irrigation water is pumped by two different groups--individual farmers and irrigation organizations. Farmers pump water from ground and surface sources, but rely heavily on groundwater sources in most of the West. Irrigation Organizations (IOs) also pump from groundwater but they rely mostly on surface water sources. Energy used to pump irrigation water by these two groups is the topic of this chapter.

The Economic Research Service, USDA, began estimating irrigation pump energy use by individual farmers in 1974 and have repeated the estimates for 1977, 1980, and 1983 (Sloggett). Comprehensive estimates of energy used for pumping water by IOs is available in the 1978 Census of IOs. Some estimates of energy used by IOs in a single state (Knutson, et al.), or group of states (King, et al.), have been made. One study attempted to estimate energy used by IOs in the 17 western states (Dvoskin, et al.) but did not include all of the organizations which pump irrigation water. Thus only partial information has been available on energy use for pumping irrigation water by IOs.

ON-FARM PUMP ENERGY USE

Procedure

Since a comprehensive measure of total energy used for irrigation pumping has not been previously attempted, this section will dwell upon the procedure used to provide this estimate. To help understand some of the problems and the complexity of estimating energy use, a brief discussion of the estimating process is in order. The calculation to estimate energy use for pumping water is:

$$E = TDH \cdot Q \cdot R$$

where:

E = energy requirement
TDH = total dynamic head is the distance (feet) water is lifted from source to field plus pressure requirements of field distribution systems including friction loss in distribution lines
Q = quantity of water pumped (acre feet)
R = energy required to lift one acre foot for one foot of elevation

While the equation is simple and the solution provides an estimate of energy used in pumping irrigation water, the process of collecting the necessary data for the equation is not simple.

Five types of energy are generally used for pumping irrigation water: electricity, diesel, gasoline, natural gas, and liquified petroleum gas (LPG). The proportion of each energy form used varies significantly among the 17 western states and among sources of irrigation water. The total pump lift (TDH) for irrigation water varies considerably among water sources and among the 17 western states. The pressure requirement component of TDH requires information on the types of distribution systems used. The quantity of water pumped (a) depends on the climate, type of distribution system used, and the kind of crops being irrigated. Estimates of energy requirements (R) to pump an acre foot of water against a unit of TDH are available (see Table 8.1) and they remain constant for given levels of pumping unit efficiency.

Background Information

A consistent source of data for estimating irrigation pump energy use by individual farmers was not available in 1974 when the demand for such information arose. A major supplemental source of data still used today is the *Irrigation Journal* which publishes an annual statistical report

Table 8.1
Fuel energy requirements for on-farm pumping one acre-foot of water at one pound per square inch (psi)

Energy Source	Horsepower Hours per unit of fuel	Fuel Units per acre-foot per psi[a]
Electricity	1.206 per kwh	4.3876
Diesel	12.35 per gallon	0.5330
Gasoline	9.875 per gallon	0.5417
Natural Gas	79 per MCF[b]	0.0677
LPG	7.9 per gallon	0.6771

Source: Delbert Schwab, agricultural engineer, Oklahoma State University, Stillwater.

[a] 60 percent pump efficiency is assumed.
[b] MCF is "1,000 cubic feet."

listing irrigation specialists in each state. These specialists are able to provide an estimate of statistics necessary to estimate on-farm irrigation pump energy use. The quality of the data provided by the specialists varies among states. Some states conduct periodic comprehensive irrigation surveys while others have collected little, if any, irrigation data. Thus, some specialists rely on surveys for their estimates while others rely only on their expertise to provide the necessary statistics. Statistics needed to estimate on-farm irrigation energy use include:

1. acres irrigated from groundwater,
2. acres irrigated from on-farm pumped surface water,
3. for each source of water:
   a. weighted average feet of lift,
   b. acreage of each irrigation distribution system used, and
   c. sources of energy used for pumping, percent of total acres,
4. average quantity of water pumped per acre irrigated.

Data Sources

Acreage Irrigated From Groundwater. Several states conduct surveys or have other procedures to estimate acreage irrigated from groundwater. Otherwise, total acreage irrigated in 1974 came from the 1974 Irrigation Survey in the *Irrigation Journal*. The proportion of state level acreage irrigated from groundwater, as published by the U.S. Geological Survey, was then multiplied by total acreage irrigated to get an estimate of acreage irrigated from groundwater in 1974 (Murray, et al.). Acreages for 1977, 1980, and 1983 were estimated by the irrigation specialists from that base.

Acreage Irrigated From Pumped Surface Water. Surface water from lakes, streams, or rivers is pumped by farmers for irrigation, but no single data source was found for acres irrigated in this manner. Therefore, each state irrigation specialist estimated the acreage irrigated with water pumped from surface sources by individual farmers.

Feet of Lift. Feet of lift is the height water must be raised from its source to the field for application. State irrigation specialists estimated a weighted statewide average feet of lift for irrigation wells and for on-farm pumped surface water. Changes in feet of lift (Appendix Table 8.1) in some of the states from 1974 through 1983 were the result of either actual changes or improved information on which the irrigation specialists based their estimates. In some cases, the changes in pump lift derived from estimated declining water levels.

Distribution System and Power Units. Information on types of distribution systems and power units in the Irrigation Journal were used, where available, for the 1974 estimates. When this information was not available for 1974 and the 1977, 1980, and 1983 surveys, irrigation specialists provided estimates of distribution system and power unit use. Distribution systems used for estimating on-farm irrigation pumping energy were big gun, center pivot, other sprinklers, and gravity. Other sprinklers include side roll, hand move, solid set, etc. Gravity systems include gated pipe, siphon tubes, flood, etc. Power sources to operate on-farm irrigation pumps include electricity, diesel, gasoline, natural gas, and liquefied petroleum gas (LPG).

Acre-Feet Applied. USGS data provide estimates of the water quantity applied per acre (Murray, et al.). Several specialists offered alternative estimates to modify USGS data.

Pumping Unit Efficiency. A new on-farm type irrigation pump has an efficiency of 75-80 percent (efficiency is a measure of energy input to water output). Efficiency declines when pump wear occurs or the pump is used under conditions for which it was not designed. Irrigation engineers in those states having significant amounts of groundwater irrigation estimated existing average operational pump efficiencies. An efficiency of 60 percent was used for estimating existing on-farm energy use (Table 8.1).

Distribution System Pressure Requirements. The following pressure requirements for various irrigation distribution systems were selected:

Type of System	Pounds per square inch (psi)
Big gun	165
Center pivot	100
Other sprinkler	70
Surface distribution	5

These data include the pressure required to overcome friction loss in lines from the pump through the distribution system and to apply water to the land. These estimates may be higher than some reported elsewhere, but frequently pressure requirements for sprinkler system operation are reported for pressure at the sprinkler nozzle but do not include pressure required to deliver water to the nozzle. Since 1977, a great deal of interest has focused on low-pressure center-pivot irrigation systems. For the 1980 and 1983 data, irrigation specialists have estimated the percentage of center pivots operating with low pressure. The average pressure required to operate the low-pressure center-pivot systems is about 35 psi. The estimated percent of low-pressure center-pivot systems for each state is shown in Appendix Table 8.9. Energy consumption estimates for 1980 and 1983 include allowances for low-pressure center-pivot systems.

During the years USDA has estimated energy use by on-farm pump irrigators. Irrigation specialists in many states have greatly improved the data necessary to make the estimates. However, these estimates do not result from comprehensive surveys and are, thus, not subject to measurable statistical error.

Survey Results

Data collected to estimate on-farm energy use for pumping irrigation water in each state are presented in Appendix Table 8.2 - 8.11. The following is a summary and discussion of the data and an estimate of energy used for on-farm pumping of irrigation water.

Area Irrigated. The area irrigated in the 17 western states by on-farm pumped water increased by 17 percent from 1974 to 1983 (Table 8.2). However that growth is slowing rapidly. From 1974 to 1977 annual growth was about one million acres, but from 1977 to 1980 it slowed to one-half million acres per year and it grew by only one-half million acres for the entire three year period between 1980 and 1983.

Groundwater was the only source of water for 75 percent of the growth and surface water was used for 10 percent of the growth. The combinations of groundwater and surface water were used for the remaining 15 percent growth in on-farm pump irrigation from 1974 to 1983. Thus, 90 percent of the increased irrigated acreage from 1974 to 1983 using on-farm pumped water relies on groundwater.

Table 8.2
Acreage irrigated with on-farm pumped water in the 17 western states

Source of Water	1974	1977	1980	1983	Change 1974-83	Change 1974-83
	------- Million Acres -------					Percent
Groundwater	22.2	25.0	25.7	25.9	3.7	17
Surface Water	5.0	5.2	5.2	5.5	0.5	10
Acreage Using Both Sources	1.8	1.9	2.6	2.6	0.8	44
TOTAL	29.0	32.1	33.5	34.0	5.0	17

"Pumped" water in this case infers that the water is lifted some elevation from its source to the level of the land to which it is applied. There is also a large and growing percentage of water delivered to farms by ditch or canal which is then pressurized and applied through some form of sprinkler system. In this sense there is a considerably larger amount of on-farm pumped water than that derived only from groundwater or surface water that must be lifted from its source to the irrigated land that it serves.

The largest changes in area irrigated with on-farm pumped water within the 17 western states between 1974 and 1983 occurred in Nebraska, Kansas, and Texas. Kansas and Nebraska together added four million acres of the total five million acre change between 1974 and 1983 while Texas declined by one million acres (Appendix Table 8.2). Nearly all of these changes were in groundwater use. Changes in irrigated acreage in the remaining 14 states were rather small during the period.

Distribution Systems. Gravity flow irrigation systems require the least amount of energy to operate and they are the most widely used systems for on-farm pumped water in the 17 western states (Table 8.3). Some gravity systems, such as siphon tubes, require energy only to pump water into ditches while others such as gated pipe require small

Table 8.3
Area irrigated with on-farm pumped water by type of distribution system, 17 western states

Distribution System	1974	1977	1980	1983	Change 1974-83	Change 1974-83
	-------- Million Acres			------		Percent
Big Gun	0.2	0.3	0.3	0.3	0.1	50
Center Pivot	3.2	5.3	6.2	6.8	3.6	113
Other Sprinkler	6.4	6.7	7.1	7.2	0.8	13
Gravity	19.2	19.8	19.9	19.7	0.5	3
TOTAL	29.0	32.1	33.5	34.0	5.0	17

amounts of pressurization. Big gun type distribution systems require the highest operating pressures and they are the least popular.

The most rapidly growing type of distribution system is the center pivot. Its use nearly doubled between 1974 and 1983 and center pivots were used for 72 percent of the growth in on-farm pump irrigation between 1974 and 1983. Nebraska and Kansas, with the largest growth in irrigated area, also had a significant portion of the growth in center pivots--2.3 million acres out of a total of 3.6 million acres (Appendix Table 8.3).

Center pivots have been the second largest energy intensive distribution systems. However, low pressure center-pivot technology became available in the late 1970s. Farmers were quick to adapt this energy saving equipment where possible; and currently about 25 percent of center pivots are low pressure (Appendix Table 8.9). News reports and informal contact with the irrigation equipment industry indicates that a high percentage of new center pivots sold are low pressure systems. Energy requirements for these systems are at or below that of other sprinkler systems.

Types of Energy Used. Electricity supplied the power for the largest share of land irrigated by on-farm pumps (Table 8.4). It also was the source of power for 3.6 million of the five million acre increase in irrigation

Table 8.4
Area irrigated (million acres) with on-farm pumped water by type of energy, 17 western states

Energy	1974	1977	1980	1983	Change 1974-83	Change 1974-83
Electricity	14.0	16.0	16.9	17.6	3.6	26
Diesel	2.0	3.4	3.4	3.5	1.5	75
Gasoline	0.3	0.2	0.3	0.2	-0.1	-33
Natural Gas	10.4	10.7	10.9	10.6	0.2	2
LPG	2.3	1.7	2.0	2.1	-0.2	-9
TOTAL	29.0	32.1	33.5	34.0	5.0	17

between 1974 and 1983. Diesel was the fastest growing source of power, but nearly all of that growth appeared between 1974 and 1977. Gasoline use for on-farm pumping of irrigation water in the West is very small and is declining. The use of LPG for on-farm pumps is also declining. Natural gas is the second most popular source of energy for on-farm irrigation pumps. While its growth has been small, some significant changes have occurred in its use. Between 1974 and 1983, natural gas use declined by nearly a million acres in Texas while Kansas and Nebraska increased their use by about 1.3 million acres (Appendix Table 8.4).

Quantity of Energy Used. The quantity of energy used for pumping (Table 8.5) is generally proportional to the irrigated acreage involved. However, recent changes in quantities of energy used have been exceeding the proportional changes in irrigated acreages because most of the increase in area irrigated has been with sprinkler distribution systems which use more energy than gravity flow systems (Table 8.3). The area irrigated with gravity flow systems remained rather steady from 1974 to 1983.

Costs of Energy Used. The cost of energy for on-farm irrigation was less than one-half billion dollars in 1974 but rose to nearly two billion dollars by 1983--a 310 percent increase (Table 8.6). This large increase in cost occurred with only a 17 percent increase in irrigated area during the same period. The obvious cause of significantly

Table 8.5
Quantity of energy used for on-farm pumped irrigation water, 17 western states

Energy	Unit	1974	1977	1980	1983	Change 1974-83	Change 1974-83
							Percent
Electricity	KWH x 10^9	15	18	19	21	6	40
Diesel	Gal x 10^6	142	246	258	264	122	86
Gasoline	Gal x 10^6	31	25	23	24	-7	-23
Natural Gas	MCF x 10^6	128	141	145	141	13	9
LPG	Gal x 10^6	198	191	213	228	30	15

Table 8.6
Total expenditures for energy for on-farm pumped water, 17 western states

Energy	1974	1977	1980	1983	Change 1974-83	Change 1974-83
	-------- Million Dollars --------					Percent
Electricity	260	489	792	965	705	271
Diesel	49	108	254	257	208	424
Gasoline	14	13	26	28	14	100
Natural Gas	97	179	364	549	452	466
LPG	53	67	125	168	115	217
TOTAL	473	856	1,561	1,937	1,464	310

higher energy costs were sharply higher energy prices (Appendix Table 8.6). Changes in prices have not been uniform among energy sources however. Between 1980 and 1983 prices of electricity, natural gas, and LPG continued their sharp rise, but diesel and gasoline prices declined in several states. The effect of these price changes and the overall trend in higher energy prices on the future of on-farm pump irrigation remain to be seen. These trends do indicate, however, why much of the irrigated sector is in serious economic jeopardy today.

IRRIGATION ORGANIZATIONS

Types of Irrigation Organizations

Irrigation specialists were not able to estimate energy used by irrigation organizations (IOs) to pump water nor were other sources for these data found when energy use data were first collected in 1974. Neither has this information been available for the 1974-1983 period. The 1978 Census of Irrigation Organizations is the latest comprehensive data source and it contains an estimate of expenditures for energy used for pumping by IOs. It is possible to estimate energy used by IOs from this expenditure data. However a discussion of background information on IOs will help to understand the extent of their pumping of irrigation water and help compare it to on-farm pumping.

The 1978 Census of Irrigation Organizations defines nine categories of IOs:

1. Unincorporated mutuals are an informal association of two or more farmers who operate irrigation supply works for their own needs.
2. Incorporated mutuals are a legally constituted association owned by irrigators to supply irrigation water.
3. Districts are a public corporation or special-purpose governmental unit which may use taxing and eminent domain powers to provide facilities for supplying irrigation water.
4. The U.S. Bureau of Reclamation (USBR) has constructed and currently operates and manages irrigation systems to supply water to other irrigation organizations and directly to farms.
5. The USBR has also constructed irrigation systems subsequently turning control of the systems over to other irrigation organizations.
6. The U.S. Bureau of Indian Affairs (USBIA) operates irrigation systems primarily on Indian reservations.

7. State and local governments operate irrigation systems in states without the district form of organizations.
8. Commercial irrigation organizations are controlled by owners rather than user-operators.
9. Other irrigation organizations include those not defined above.

There were 7,034 IOs in the 17 western states in 1978 irrigating 21.3 million acres (Table 8.7). The small, informal, unincorporated mutuals represent half of all IOs but they contributed only 8 percent of all water delivered to farms and ranches. Irrigation districts represent only 11 percent of all IOs but supplied 50 percent of the water delivered to farms and ranches by IOs.

The USBR constructed and operated IOs represent only 1 percent of all IOs but they conveyed 29 percent of the water. However, they delivered only 2 percent of the water to farms and ranches (Table 8.7). The 294 USBR constructed but <u>user</u> operated IOs are not included separately in the 7,034 total but are displayed in the table to indicate how many of the facilities in the other eight categories of IOs were constructed by USBR. That is, some of the irrigation districts, incorporated mutuals, and other categories include USBR constructed facilities. The USBR estimates that facilities constructed by their organization delivered 24.3 million acre feet (maf) to farms and ranches in 1978 (Water and Land Resource Accomplishments, 1978). Thus, USBR constructed facilities were involved in supplying (24.3 ÷ 62) 39 percent of all irrigation water delivered to farms and ranches in the 17 western states by IOs in 1978. State and local government irrigation organizations supplied only 2 percent of farm and ranch irrigation water so nearly 60 percent of the irrigation water delivered to farms and ranches was delivered by non governmental irrigation organizations.

Quantity of water conveyed and quantity of water delivered to farms and ranches as displayed in Table 8.7 need some explanation. Quantity of water delivered to farms and ranches is what IOs actually deliver to farms for irrigation. An accounting of water conveyed by IOs is given in Table 8.8. Conveyance loss is the quantity of water lost to evaporation and seepage from IO reservoirs, canals, and ditches before it gets to the water user. Other releases include quantity of water which was spilled or wasted from the conveyance system without being

Table 8.7
Water conveyed and delivered by irrigation organizations in the 17 western states, 1978

Irrigation Organization	Number of Organization		Quantity of Water Conveyed		Water Delivered to Farm & Ranches		Area Irrigated	
	Number (percent)		Million Acre-Feet —(percent)—				Million Acres —(percent)—	
Uninc. Mutual	3498	(50)	7	(4)	5	(8)	2.0	(9)
Inc. Mutual	2409	(34)	33	(20)	20	(32)	7.0	(33)
District	795	(11)	59	(35)	31	(50)	10.8	(51)
USBR Constructed								
USBR Operated	63	(1)	48	(29)	1	(2)	.3	(1)
User Operated[a]	294[b]				24[b]	(39)[b]	6.3[b]	
USBI	133	(2)	4	(2)	2	(3)	.6	(3)
St. & Loc. Gov't	67	(1)	6	(4)	1	(2)	.3	(1)
Commercial	52	(1)	1	(1)	1	(1)	.2	(1)
Other	17	(*)	9	(5)	1	(1)	.2	(1)
17 Western States	7034	(100)	167	(100)	62	(100)	21.3	(100)

*Less than one percent

[a]User operated USBR constructed are listed to show how many of the mutual and district irrigation organizations facilities were constructed by USBR.

[b]These numbers are embodied in the data for other categories of IO.

delivered to any user or that used for power production. Water delivered to farms and ranches is the quantity delivered to the farm gate. The 48.6 million acre feet (maf) conveyed to other IOs is dispersed among farmers and ranchers, MIR, conveyance loss, and other releases. That is, part of the 48.6 maf delivered to other IOs is included in the 61.9 maf delivered to farmers, 5.5 maf delivered for municipal, industrial and recreational uses. Thus only

Table 8.8
An accounting of water conveyed by irrigation organizations

Item	Million Acre Feet
Municipal, Industrial, Recreational (MIR)	5.5
Conveyance Loss	18.8
Other Releases	32.2
To Other IOs	48.6
To Farms and Ranches	61.9
TOTAL	167.0

(167 - 48.6) 118.4 maf are actually distributed to water users by IOs. This method of water accounting is rather cumbersome, but necessary, to account for water distributed from one IO to another.

Agricultural Water Pumped by IOs

This chapter is concerned with energy used for pumping irrigation water. Thus, the quantity of water of concern is that portion of the 118.4 maf distributed by IOs that could be related to irrigation and is also pumped. Conveyance loss and other releases represent (18.8 + 32.2) 51 maf of the 118.4 maf of water distributed by IOs. If it is assumed that the other two categories of delivery, MIR and farms and ranches, share proportionately in the conveyance losses and other releases then [(5.5/61.9) x (51) + 5.5] 10 maf can be allocated to MIR leaving (118.4 - 10) 108.4 maf that are related to irrigation. However, only a portion of the 108.4 maf are pumped. Some IOs deliver water to farm and ranches entirely by gravity. The Census of irrigation organizations provides only a partial answer to the question of how much is pumped and how much is delivered by gravity.

The 1978 Census of Irrigation Organizations provides data on the quantity of water pumped from groundwater, but it does not distinguish between pumped and non-pumped surface water handled by IOs. The remainder of this section explains the procedure used to estimate water pumped and energy used by IOs for agricultural purposes. A reader may skip to the next major section without loss of understanding for the rest of the chapter.

Groundwater. The quantity of water pumped by IOs from groundwater is shown in Table 8.9. Pumped groundwater supplied only about (2.7/108.4) 2.5 percent of the agricultural water distributed by IOs in 1978. Groundwater from flowing (artesian) wells provided less than 0.01 percent. Thus surface water is the major source for IOs but the amount pumped remains to be determined.

Surface Water. The 1978 Census of Irrigation Organizations defined two types of surface water pumps, diversion pumps and "other" pumps, and provides information on pump capacities and pumping lift but not quantity of water pumped (Table 8.10). Diversion pumps are used to divert water from streams, lakes, reservoirs, and ponds into water conveyance pipelines, ditches, or canals. Only 548 of the 7,034 IOs in the 17 western states used diversion pumps. The remaining 6,486 IOs either get their water from other IOs with diversion pumps, wells, or by gravity from reservoirs or diversion dams. Some may even get water from all of the above sources. Three states--California, Texas, and Washington--account for over 80 percent of diversion pumping capacity.

"Other" pumps are defined as pumps within an IO system designed to relift water from one conveyance system to another or to pressurize water for distribution systems. "Other" pumps also include those used for drainage. Pumps defined as "other" are used by 500 of the 7,034 IOs. Four states--Arizona, California, Texas, and Washington--account for over 90 percent of "other" pumping capacity.

It is not possible to add all IOs using one of the three types of pumps and get a total number of IOs using pumps. The sum of IOs using at least one of the three types of pumps is (522 + 548 + 500) 1,570. Thus it is possible to say that at least 5,464 or 78 percent of the IOs in the 17 western states do not use pumps within their delivery systems. How many of the 5,464 IOs that do not use pumps but receive water that has been pumped by other IOs is not known.

Published data that is useful in estimating the quantity of water pumped by IOs is the capacity of IO pumps (Table 8.10). If all diversion pumps were operated at capacity--163,800 acre feet per day (Table 8.10)--for 365 days they could pump 59.8 million acre feet (1 cubic

Table 8.9
Selected data for irrigation organizations supplying water from pumped wells

State	Organizations	Pumped Wells	Average Pump Lift	Acre-Feet Delivered
	number	number	Feet	1,000
Arizona	60	1000	244	664[a]
California	220	1834	137	815
Colorado	36	106	51	98[a]
Idaho	41	428	154	523[a]
Kansas	0	0	0	0
Montana	6	9	21	17
Nebraska	2	b	87	10
Nevada	2	b	101	16[a]
New Mexico	28	83	55	46[a]
North Dakota	3	b	87	2
Oklahoma	0	0	0	0
Oregon	12	19	90	56[a]
South Dakota	b	b	b	3
Texas	4	18	213	12
Utah	82	141	129	115[a]
Washington	18	90	190	316
Wyoming	8	22	116	12[a]
17 Western States	522	3,783	161	2,705[a]

Source: Table 16.

[a]Adjusted for quantity of ground water delivered from artesian flowing wells.
[b]Not available.

foot per second is equivalent to about 2.0 acre feet per day). This is the upper limit on the annual quantity of water that could be diverted from surface water. "Other" pumps, the pumps used within IOs, have a total capacity of 248,400 acre feet per day or 90.6 million acre feet per year.

Table 8.10
Surface water pumps used by irrigation organizations in the 17 western states

States	Diversion Pumps				Other Pumps			
	Organi-zations	Pumps	Total Pump Capacity	Average Lift	Organi-zations	Pumps	Total Pump Capacity	Average Lift
	no.	no.	1000 cfs	feet	no.	no.	1000 cfs	feet
Arizona	14	22	0.70	59	13	265	20.20	45
California	112	707	35.00	45	195	418	74.60	91
Colorado	28	52	1.00	54	30	85	1.20	59
Idaho	84	339	4.10	187	50	388	3.40	39
Kansas	a	a	a	a	a	a	a	a
Montana	57	173	3.40	45	35	107	2.40	20
Nebraska	5	7	0.01	20	6	a	0.30	24
Nevada	2	a	a	13	2	a	a	6
New Mexico	13	20	0.03	22	9	26	0.08	30
North Dakota	5	9	0.30	24	1	a	a	20
Oklahoma	a	a	a	a	a	a	a	a
Oregon	68	260	4.30	81	46	199	1.70	159
South Dakota	7	25	0.07	152	1	a	a	50
Texas	46	215	16.60	25	42	381	8.40	15
Utah	41	73	0.40	52	13	50	1.40	204
Washington	39	186	15.70	221	30	721	10.90	79
Wyoming	26	74	0.30	51	24	70	0.30	35
17 Western States	548	2168	81.90	82	500	6576	124.20	79

[a] Not available.

A number of variables affect the length of time any particular IO may be required to pump water including: length of the growing season, seasonal rainfall, and crops grown. Use of the pumped water for supplemental or primary irrigation will also affect pump operating time. A number of IOs, including some USBR offices, in the 17 western states were contacted to find out how much irrigation water they pumped in a typical year and what their pumping capacity was. From that information, it was possible to estimate how many days of pumping at full capacity it would take to pump all the water they pumped in a typical year. Results of that informal survey are presented in Table 8.11. The actual pumping season for IOs is longer than the number of days listed in Table 8.11 because IOs do not pump at full capacity throughout the entire irrigation season.

The 17 western states were divided into six areas that are similar with respect to irrigation season. The full capacity operating times in Table 8.11 were used uniformly

Table 8.11
Typical full capacity operating times for irrigation organization pumps in the 17 western states

Area	Days Operated at Full Capacity	
	Diversion Pumps	Other Pumps
Arizona, New Mexico	200	175
California	160	150
Idaho, Oregon	120	110
Washington	[a]	[a]
Montana, Colorado	60	60
Utah, Nevada	[a]	[a]
Wyoming	[a]	[a]
N. and S. Dakota	70	60
Nebraska, Kansas	[a]	[a]
Oklahoma	[a]	[a]
Texas	100	90

[a]Unknown.

for each state in the six regions for purposes of estimating the quantity of surface water pumped. Diversion pumps in some areas were estimated to operate longer than other pumps because some IOs in those areas often pump water into storage reservoirs outside the growing season. Other pumps typically operate only during the growing season.

The quantity of surface water pumped by IOs can be estimated by multiplying days of full capacity pumping (Table 8.11) times daily pumping capacity (Table 8.10). The quantity of water pumped out of rivers and reservoirs by diversion pumps in a typical year was estimated at 21.3 maf. "Other" pumps were estimated to pump 35.2 maf annually (Table 8.12).

If all diversion pumped water was re-pumped by "other" pumps then it could be concluded that (35.2/105.9) 33.2 percent of the irrigation water conveyed by IOs required some pumping. If none of the diverted water was re-pumped it could be concluded that [(21.3 + 35.2)/105.9] 53.4 percent of the irrigation water conveyed by IOs required pumping. However, the quantity of water diverted into canals and/or reservoirs and then re-pumped by "other" pumps is not known. Some large IOs such as the California Central Valley Project and the Washington Columbia Basin Project divert water with pumps into canals or reservoirs and then re-pump that same water with "other" pumps, sometimes the water is re-pumped more than once. Thus, it may only be concluded that between 33.2 and 53.4 percent of the surface irrigation water conveyed by IOs requires pumping.

Pumping Energy Used by Irrigation Organizations

The amount of energy required by IOs to pump 2.5 maf feet of groundwater and 56.5 maf of surface water by diversion and "other" pumps may be estimated from energy expenditure data provided by the Census of IOs (Table 8.13). Combustion fuel prices were relatively uniform across the 17 western states in 1977 and 1978 but electric rates for pumping irrigation water were highly variable (Miller and Schluntz). Non USBR electric rates ranged from $.011 per kwh in the Pacific Northwest to $.045 per kwh in the Great Plains (Appendix Table 8.8). With the heavy concentration of pumping in California and the Pacific Northwest and with USBR being a large supplier of power for pumping, a 1978 weighted average electric rate for pumping

Table 8.12
Estimated quantity of surface water pumped by irrigation organizations in the 17 western states, 1978

State	Diversion Pumps	Other Pumps
	------ 1,000 Acre Feet ----	
Arizona	280	7,070
California	11,200	22,380
Colorado	120	144
Idaho	984	748
Kansas	a	a
Montana	408	288
Nebraska	1	36
Nevada	a	a
New Mexico	12	28
North Dakota	42	a
Oklahoma	a	a
Oregon	1,032	374
South Dakota	9	a
Texas	3,320	1,512
Utah	48	171
Washington	3,768	2,398
Wyoming	36	36
17 Western States	21,260	35,185

[a]Not available.

irrigation water in the 17 western states falls in the range of $.0075 to $.015 per kwh. The estimated quantity of energy used by IOs in the 17 western states is shown in Table 8.13.

SUMMARY OF ENERGY USED FOR PUMPING

The quantity of energy used by on-farm pumps is much larger than that used for pumping by IOs as shown in Table

Table 8.13
Energy used by IOs for pumping irrigation water in the 17 western states, 1978

Energy	Unit	Quantity of Energy	Price/Unit (dollar)	Value ($1,000)
Electricity	KWH x 10^9	2.6-5.1	38,278	.0075-.015
Diesel	Gal x 10^6	1.4	662	.46
Gasoline	Gal x 10^6	2.2	1,345	.60
Natural Gas	MCF x 10^6	3.0	4,537	1.50
LPG	Gal x 10^6	0.3	123	.40

Source: 1978 Census of Irrigation Organizations, United States Department of Agriculture, Agricultural Statistics, 1979, and Correspondence with United State Bureau of Reclamation, Denver, Colorado.

8.14. Very small amounts of diesel, gasoline, natural gas, and LPG are used by IOs for pumping irrigation water because most of their water is pumped by very large electric motors. On-farm irrigation water pumps are typically much smaller where standard internal combustion engines are applicable.

The larger area irrigated by on-farm pumps explains part of the reason for higher on-farm energy use but differences in the source of pumped water and the types of distribution systems between the two groups also affect energy use. Groundwater is the predominant source of water for on-farm pumping and typically requires higher pumping lifts than for surface water--the predominant source of water for IOs. Sprinkler systems require much more energy for operation than do surface distribution systems. In 1977-78, on-farm pumped water was distributed by sprinkler systems on about 12 million acres (Table 8.3) compared to about 3.6 million acres of sprinkler supplied water by IO pumps (Appendix Table 8.10).

A major portion of the irrigated area in the 17 western states requires energy to pump the water, though the exact acreage is unknown. It is possible to compare

Table 8.14
Estimated total energy used per year for irrigation pumping[a]

Item	Unit	Quantity	
		On-Farm	IO
Electricity	kwh x 10^9	18	2.6-5.1
Diesel	Gal x 10^6	246	1.4
Gasoline	Gal x 10^6	25	2.2
Natural Gas	mcf x 10^6	141	3.0
LPG	Gal x 10^6	191	0.3

[a]On-farm data for 1977 and IO data for 1978.

the quantity of water delivered to farms and ranches with, and without, the aid of pumps.

In 1977 on-farm pumps provided 72.7 million acre feet of water for irrigation (Appendix Table 8.7). Irrigation organizations conveyed 108.4 maf for irrigation in 1978 of which an estimated 35 to 57 maf was pumped. Thus, total water conveyed for irrigation in the West in the 1977-1978 period was (72.7 + 108.4) 181.1 maf annually. However, conveyance losses by IOs must be considered to compare on-farm water pumped and IO pumped water delivered to farms and ranches.

Conveyance losses for on-farm pumped water is essentially zero because no off-farm delivery system is involved. However IO pumped water is conveyed over long distances and sometimes stored in reservoirs for some length of time. About 43 percent (61.9 maf/108.4 maf) of the agricultural water conveyed by IOs was lost to conveyance before it was delivered to farms and ranches for irrigation in 1978. Thus, only about [72.7 maf + (108.4 maf x .57)] 134.5 maf were actually available for irrigation.

If the 43 percent conveyance loss is applied to the 35 to 57 maf of water pumped by IOs, then only 20 to 32 maf of IO pumped water was actually delivered to farms and ranches. Thus an estimated total of (20 + 72.7) 92.7 to (32 + 72.7) 104.7 maf of pumped water was delivered annually to farms and ranches in the 1977-78 period. Comparing the 134.5 maf of water available to farms and ranches for irrigation with the 92.7 to 104.7 maf pumped, it may be concluded that between 69 and 78 percent of the water available for irrigation in the 17 western states requires energy using pumps.

Quantity of water pumped and area irrigated are not proportional, however, because application rates per acre differ among the 17 western states (Appendix Table 8.1). In the far West, application rates typically range from 2-4 acre feet per acre while in the Great Plains, application rates are generally from 1-2 acre feet per acre. In the Great Plains, where on-farm pumped water is dominant, an equivalent amount of water could irrigate twice as much land as in the far West where IO pumped water is dominant. Thus, the percent of area irrigated with pumped water in the 17 western states is higher than the percent of the water pumped. While 69 to 78 percent of the water available for irrigation is pumped perhaps as much as 80 to 90 percent of the area irrigated is done so with the aid of energy using pumps.

The following appendix tables provide the most recent and comprehensive estimates of energy use and water use for irrigation in the 17 western states.

REFERENCES

Dvoskin, Dan and others. "Energy Use for Irrigation in the Seventeen Western States." CARD Report, Iowa State University, Ames, Iowa, July 1975.

"Irrigation Journal." Brantwood Publication Inc., Elm Grove, Wisconsin, November-December 1974.

King, Larry and others. "Energy and Water Consumption of Pacific Northwest Systems." BNWL-RAP-19-UC-11, Department of Agricultural Engineering, Oregon State University, Corvallis, August 1978.

Knutson, G.D. and others. "Pumping Energy Requirements for Irrigation in California." Special Publication No. 3215. Division of Agricultural Science, Univ. of California, Davis, July 1977.

Miller, Joe and Larry Schluntz. Personal communication, Division of Water and Land Technical Services, U.S. Bureau of Reclamation, Denver Federal Center, Denver, Colorado.

Murray, C. Richard and others. "Estimated Use of Water in the U.S. in 1970." USGS Circular 676, U.S. Department of Interior, U.S. Geological Survey, 1972.

Sloggett, Gordon R. "Energy and U.S. Agriculture: Irrigation Pumping 1974-80." AER 495, U.S. Department of Agriculture, Economic Research Service, December 1982.

"Water and Land Resource Accomplishments, 1978." Statistical Appendix I, U.S. Department of Interior, Water and Power Resources Service, 1979.

Appendix 8.A: Data Related to Energy Use for Pumping Irrigation Water

Appendix Table 8.1
Weighted average feet of lift required for on-farm pumping and acre-feet of irrigation water applied by source for the 17 western states

Region/ State	Groundwater				Surface Water				Acre-Feet Applied
	1974	1977	1980	1983	1974	1977	1980	1983	
	------------- Feet of Lift -------------								
Arizona	350	375	400	425	0	0	0	0	5.40
California	110	110	135	140	10	10	10	10	3.17
Colorado	115	120	125	125	10	10	10	10	1.10
Idaho	266	266	225	280	0	11	12	13	3.20
Kansas	180	180	190	190	15	15	15	15	1.70
Montana	100	100	100	100	60	60	60	60	2.70
Nebraska	100	100	100	100	20	20	20	20	1.75
Nevada	100	100	100	100	20	20	20	20	5.00
New Mexico	250	250	260	265	5	5	5	5	2.75
North Dakota	75	75	75	75	35	35	35	35	1.00
Oklahoma	200	200	200	200	20	16	16	16	1.83
Oregon	266	266	266	266	11	11	11	11	3.00
South Dakota	70	80	120	120	150	150	130	130	1.25
Texas	200	200	210	215	40	40	40	40	1.50
Utah	225	225	225	225	15	15	15	15	2.75
Washington	287	287	270		26	26	26		4.20
Wyoming	150	150	150	150	25	25	25	25	1.83

Appendix Table 8.2
Acreage irrigated with on-farm pumped water, by source in the 17 western states

Region/State	Groundwater				Surface Water			
	1974	1977	1980	1983	1974	1977	1980	1983
	1,000 Acres							
Arizona	552	550	550	548	0	0	0	0
California	4,073	4,388	4,065	4,065	380	410	410	410
Colorado	900	940	940	940	45	50	60	60
Idaho	1,056	1,200	1,250	1,270	478	482	527	537
Kansas	2,230	3,073	3,489	3,489	65	75	85	85
Montana	40	57	58	60	284	316	389	401
Nebraska	4,074	5,670	6,316	6,850	505	440	440	440
Nevada	170	170	170	170	34	34	34	34
New Mexico	634	585	653	633	43	43	43	45
North Dakota	33	85	127	141	23	11	11	15
Oklahoma	680	730	746	645	40	118	120	120
Oregon	246	264	292	328	644	686	738	811
South Dakota	43	149	198	210	91	150	174	174
Texas	7,090	6,718	6,345	5,973	1,451	1,451	1,095	1,073
Utah	60	60	70	70	164	164	80	80
Washington	242	260	330	378	701	749	930	1,142
Wyoming	125	125	130	130	50	50	75	79
TOTALS	22,246	25,024	25,729	25,902	4,998	5,229	5,211	5,506

(Continued)

Appendix Table 8.2 (Cont.)

	Ground and Surface Water				Total			
Region/State	1974	1977	1980	1983	1974	1977	1980	1983
	1,000 Acres							
Arizona	391	390	390	390	943	940	940	938
California	0	0	200	200	4,453	4,798	4,675	4,675
Colorado	700	710	720	720	1,645	1,700	1,720	1,720
Idaho	100	150	180	180	1,634	1,832	1,957	1,987
Kansas	10	10	15	15	2,305	3,158	3,589	3,589
Montana	0	0	0	0	324	373	447	461
Nebraska	176	175	175	175	4,755	6,285	6,931	7.465
Nevada	0	0	0	0	204	204	204	204
New Mexico	143	175	180	172	820	803	876	850
North Dakota	0	0	0	0	56	96	138	156
Oklahoma	0	0	0	0	720	848	866	765
Oregon	0	0	0	0	890	950	1,030	1,139
South Dakota	0	0	0	0	134	299	372	384
Texas	256	256	712	712	8,797	8,425	8,152	7,758
Utah	0	0	0	0	224	224	150	150
Washington	0	0	0	0	943	1,009	1,260	1,520
Wyoming	50	50	50	53	225	225	255	264
TOTALS	1,826	1,916	2,622	2,617	29,070	33,041	33,562	34,025

Appendix Table 8.3
Acreage irrigated with on-farm pumped water in the 17 western states, by type of distribution system

Region/State	Big Gun				Center Pivot			
	1974	1977	1980	1983	1974	1977	1980	1983
	1,000 Acres							
Arizona	0	0	0	0	28	28	28	30
California	0	0	0	0	0	0	0	0
Colorado	0	0	0	0	353	546	600	633
Idaho	0	0	0	2	179	210	225	224
Kansas	45	13	52	52	449	806	987	987
Montana	1	5	11	12	23	36	60	66
Nebraska	0	63	69	70	1,025	2,136	2,356	2,815
Nevada	0	0	0	0	10	10	10	10
New Mexico	0	0	0	0	233	129	224	246
North Dakota	3	13	11	11	26	74	117	137
Oklahoma	20	15	22	29	81	124	208	155
Oregon	0	28	38	46	97	104	155	233
South Dakota	12	27	27	27	34	159	226	239
Texas	88	84	81	78	439	589	570	543
Utah	0	0	0	0	19	19	11	11
Washington	0	10	16	4	198	211	389	410
Wyoming	0	0	1	2	70	70	85	93
TOTALS	169	258	328	333	3,264	5,251	6,248	6,832

(Continued)

Appendix Table 8.3 (Cont.)

Region/State	Other Sprinkler				Surface			
	1974	1977	1980	1983	1974	1977	1980	1983
	1,000 Acres							
Arizona	37	28	28	28	876	883	883	879
California	935	1,007	981	981	3,517	3,790	3,693	3,693
Colorado	64	50	51	57	1,227	1,103	1,068	1,030
Idaho	1,209	1,400	1,530	1,581	224	220	200	181
Kansas	45	64	75	75	1,764	2,273	2,473	2,473
Montana	54	79	112	116	240	252	263	267
Nebraska	467	565	623	635	3,263	3,519	3,881	3,946
Nevada	10	10	10	11	184	182	182	183
New Mexico	16	30	59	68	571	642	591	536
North Dakota	2	3	5	4	23	5	5	4
Oklahoma	270	349	364	265	347	360	269	316
Oregon	792	817	832	860	0	0	3	0
South Dakota	65	88	93	94	22	23	24	24
Texas	1,495	1,356	1,304	1,242	6,774	6,402	6,196	5,896
Utah	179	179	120	120	24	24	18	19
Washington	707	756	845	957	37	30	9	150
Wyoming	88	88	86	89	66	66	81	81
TOTALS	6,435	6,869	7,118	7,183	19,180	19,758	19,863	19,678

Appendix Table 8.4

Acreage irrigated with on-farm pumped water in the 17 western states, by type of energy

Region/State	Electricity				Diesel				Gasoline			
	1974	1977	1980	1983	1974	1977	1980	1983	1974	1977	1980	1983
	1,000 Acres											
Arizona	612	648	648	648	0	0	0	0	0	0	0	0
California	4,364	4,757	4,461	4,460	3	9	127	130	0	0	0	0
Colorado	1,100	1,138	1,135	1,163	100	100	49	100	20	20	100	21
Idaho	1,568	1,750	1,876	1,901	49	60	60	62	16	14	13	15
Kansas	169	503	787	787	138	534	460	460	22	0	0	0
Montana	270	315	382	395	36	39	46	48	13	14	13	13
Nebraska	1,308	1,885	2,148	2,448	1,360	2,262	2,217	2,260	118	62	69	74
Nevada	159	163	163	159	40	40	40	41	2	0	0	2
New Mexico	203	223	232	245	46	76	108	100	31	22	8	10
North Dakota	42	80	122	132	8	13	12	23	3	2	2	0
Oklahoma	102	141	140	133	48	58	73	71	20	21	20	16
Oregon	890	950	1,026	1,139	0	0	3	0	0	0	0	0
South Dakota	52	143	215	227	36	103	101	103	8	7	7	7
Texas	1,904	1,997	1,914	1,824	102	97	92	88	87	83	81	78
Utah	190	190	130	130	15	15	8	10	0	0	0	0
Washington	943	1,009	1,256	1,519	0	0	3	0	0	0	0	0
Wyoming	191	191	219	231	18	18	20	19	3	3	3	2
TOTAL	14,067	16,083	16,854	17,541	1,993	3,424	3,414	3,515	343	248	314	238

(Continued)

Appendix Table 8.4 (Cont.)

Region/State	Natural Gas				Liquefied Petroleum Gas			
	1974	1977	1980	1983	1974	1977	1980	1983
	1,000 Acres							
Arizona	330	291	291	290	0	0	0	0
California	85	31	85	85	0	0	0	0
Colorado	330	332	332	335	100	100	100	101
Idaho	5	6	6	7	1	2	2	2
Kansas	1,792	1,911	2,067	2,067	183	208	273	273
Montana	0	0	0	0	3	4	5	5
Nebraska	637	1,319	1,524	1,570	1,331	754	970	1,113
Nevada	0	0	0	0	2	0	0	2
New Mexico	484	473	451	435	54	7	74	60
North Dakota	0	0	1	1	1	0	0	0
Oklahoma	435	472	498	413	113	152	132	132
Oregon	0	0	0	0	0	0	0	0
South Dakota	0	0	0	0	37	44	47	47
Texas	6,307	5,868	5,705	5,424	395	376	358	342
Utah	0	0	0	0	17	17	11	9
Washington	0	0	0	0	0	0	0	0
Wyoming	7	7	8	8	6	6	4	4
TOTAL	10,407	10,704	10,962	10,635	2,242	1,668	1,974	2,090

Appendix Table 8.5
Quantity of energy used for on-farm pumped irrigation water for the 17 western states

Region/State	Electricity				Diesel				Gasoline			
	1974	1977	1980	1983	1974	1977	1980	1983	1974	1977	1980	1983
Groundwater:	Million kwh				1,000 Gallons							
Arizona	1,736	2,639	2,802	2,963	0	0	0	0	0	0	0	0
California	3,432	3,706	3,594	3,762	0	0	15,712	15,712	0	0	0	0
Colorado	331	343	412	475	4,243	4,243	4,062	4,243	300	300	303	303
Idaho	2,750	3,282	3,555	3,586	8,262	10,028	10,184	11,205	3,414	3,136	2,866	3,463
Kansas	109	164	190	199	7,747	39,578	31,583	31,583	1,938	0	0	0
Montana	28	62	64	64	591	1,061	1,089	1,094	687	919	943	855
Nebraska	772	1,181	1,321	1,582	84,140	139,913	134,634	143,197	3,289	4,862	5,264	5,915
Nevada	310	166	166	174	7,862	4,122	4,122	4,122	492	0	0	0
New Mexico	332	289	327	359	9,383	11,922	18,279	17,492	7,825	4,474	1,759	2,187
North Dakota	14	43	63	47	302	714	637	813	142	127	177	0
Oklahoma	90	130	147	120	3,563	3,034	4,281	3,470	2,229	1,898	2,142	1,240
Oregon	265	285	316	350	0	0	731	0	0	0	0	0
South Dakota	10	56	100	112	732	3,578	3,902	4,005	204	131	195	100
Texas	1,056	1,102	1,224	1,122	5,220	4,945	5,224	5,046	6,530	6,186	6,536	6,314
Utah	112	112	108	113	908	908	744	744	0	0	0	0
Washington	959	1,021	1,276	1,538	0	0	1,202	0	0	0	0	0
Wyoming	167	167	167	173	1,561	1,558	1,969	1,832	244	243	246	134
TOTAL GROUNDWATER	12,473	14,748	15,832	16,739	134,514	225,604	238,355	244,558	27,294	22,276	20,431	20,511

(Continued)

Appendix Table 8.5 (Cont.)

Region/State	Electricity 1974	1977	1980	1983	Diesel 1974	1977	1980	1983	Gasoline 1974	1977	1980	1983
Surface Water:												
Arizona	0	0	0	0	0	0	0	0	0	0	0	0
California	222	239	239	239	69	198	0	0	0	0	0	0
Colorado	2	2	4	3	0	0	0	0	0	0	10	6
Idaho	451	456	496	518	1,524	1,965	1,965	1,801	595	491	491	501
Kansas	1	2	2	2	274	1,440	1,557	1,557	0	0	0	215
Montana	118	153	231	252	1,513	1,751	2,622	2,857	516	657	656	714
Nebraska	9	54	54	54	0	6,451	5,499	5,515	1,126	224	215	197
Nevada	8	14	14	14	197	347	347	347	12	0	0	0
New Mexico	4	4	11	12	0	0	0	0	0	0	0	0
North Dakota	2	3	4	3	32	71	42	59	15	0	0	0
Oklahoma	0	11	8	3	360	1,769	2,178	2,431	0	531	454	507
Oregon	445	473	510	886	0	0	0	0	0	0	0	0
South Dakota	26	51	68	74	1,786	4,100	3,945	4,360	497	586	494	617
Texas	116	128	95	95	762	766	569	554	477	479	356	363
Utah	146	146	65	68	1,189	1,189	365	365	0	0	0	0
Washington	1,136	1,209	1,511	1,821	0	0	0	0	0	0	0	0
Wyoming	6	6	15	16	56	56	49	36	18	18	20	27
TOTAL SURFACE WATER	2,692	2,951	3,327	4,071	7,762	20,004	19,138	19,882	3,226	2,986	2,606	3,147
TOTAL GROUND AND SURFACE WATER	15,156	17,699	19,159	20,810	142,276	245,608	257,493	264,440	30,520	25,262	23,037	23,658

(Continued)

Appendix Table 8.5 (Cont.)

Region/State	Natural Gas				Liquefied Petroleum Gas			
	1974	1977	1980	1983	1974	1977	1980	1983
	1,000 MCF				1,000 Gallons			
Groundwater:								
Arizona	14,432	18,304	19,435	20,457	0	0	0	0
California	1,000	1,136	1,638	1,638	0	0	0	0
Colorado	2,050	2,072	2,072	2,222	1,842	1,842	1,864	1,864
Idaho	116	117	118	159	387	390	392	397
Kansas	19,383	23,982	24,282	24,282	16,961	23,208	28,809	28,809
Montana	7	0	0	0	66	255	262	213
Nebraska	6,166	12,762	14,474	15,545	94,569	72,929	92,108	111,011
Nevada	0	0	0	0	615	0	0	0
New Mexico	15,161	11,559	11,873	12,165	17,117	18,640	19,789	16,412
North Dakota	0	0	0	0	59	0	0	0
Oklahoma	5,943	7,473	8,703	6,201	11,143	16,608	14,728	12,403
Oregon	0	0	0	0	0	0	0	0
South Dakota	0	0	0	0	1,187	1,974	2,441	2,505
Texas	60,401	59,850	58,822	56,694	32,649	34,951	32,677	31,455
Utah	0	0	0	0	1,624	1,624	1,551	1,551
Washington	0	0	0	0	0	0	0	0
Wyoming	122	121	92	84	610	609	616	505
TOTAL GROUND WATER	124,781	137,376	141,509	136,498	178,219	177,030	195,237	207,125

(Continued)

Appendix Table 8.5 (Cont.)

	Natural Gas				Liquefied Petroleum Gas			
Region/State	1974	1977	1980	1983	1974	1977	1980	1983
Surface Water:								
Arizona	0	0	0	0	0	0	0	0
California	0	0	0	0	0	0	0	0
Colorado	0	0	0	0	0	0	12	8
Idaho	29	46	46	47	0	148	148	156
Kansas	0	0	0	591	428	1,310	1,709	3,762
Montana	0	0	0	0	215	273	410	446
Nebraska	0	588	591	603	6,567	3,362	3,762	3,820
Nevada	0	0	0	0	15	0	0	0
New Mexico	4	3	9	13	0	0	0	0
North Dakota	0	0	0	5	6	0	0	0
Oklahoma	0	121	125	137	2,251	4,759	4,882	5,492
Oregon	0	0	0	0	0	0	0	0
South Dakota	0	0	0	0	2,896	3,297	3,495	3,473
Texas	3,574	3,414	2,534	2,523	4,169	4,193	3,111	3,098
Utah	0	0	0	0	2,126	2,126	799	799
Washington	0	0	—	0	0	0	0	0
Wyoming	0	0	10	10	55	55	25	34
TOTAL SURFACE WATER	3,603	4,172	3,315	3,929	18,722	19,523	18,353	21,088
TOTAL GROUND AND SURFACE WATER	128,384	141,548	144,824	140,427	196,941	196,553	213,590	228,213

Appendix Table 8.6
Total cost of energy for on-farm pumped irrigation water, by region and state

Region/State	Electricity				Diesel				Gasoline			
	1974	1977	1980	1983	1974	1977	1980	1983	1974	1977	1980	1983
	Million Dollars											
Groundwater:												
Arizona	34.7	55.5	126.1	133.3	0.0	0.0	0.0	0.0	0.0	0.0	0.0	0.0
California	68.6	155.3	215.8	233.2	0.0	0.0	15.2	15.2	0.0	0.0	0.0	0.0
Colorado	7.3	11.9	20.7	35.6	1.5	1.9	4.1	4.0	.1	.1	.3	.4
Idaho	44.0	65.6	106.6	107.6	3.1	4.1	9.9	10.7	1.6	1.6	3.4	4.2
Kansas	2.2	5.8	10.3	12.9	2.8	17.0	30.9	30.6	.9	0.0	0.0	0.0
Montana	.5	10.1	2.5	3.2	.2	.4	1.1	1.1	.3	.4	1.1	1.1
Nebraska	15.4	53.1	66.1	102.8	28.6	62.9	133.2	138.9	1.6	2.4	6.2	7.5
Nevada	4.6	5.0	8.3	10.8	3.1	1.8	4.3	4.3	.2	0.0	0.0	0.0
New Mexico	6.6	8.7	16.3	27.3	3.3	4.6	18.0	17.4	3.6	2.2	1.9	2.6
North Dakota	.3	1.6	2.4	3.1	.1	.3	.6	.8	[b]	.1	.2	0.0
Oklahoma	2.0	3.9	6.6	8.4	1.2	1.3	4.3	3.5	1.0	1.0	2.5	1.4
Oregon	3.5	3.9	9.5	14.7	.1	0.0	.7	0.0	0.0	0.0	0.0	0.0
South Dakota	.2	2.3	4.6	5.6	.2	1.5	4.0	3.9	.1	.1	.2	.1
Texas	23.2	33.1	52.4	72.9	1.7	2.2	5.2	4.7	2.8	3.4	7.4	7.0
Utah	1.7	2.8	3.7	5.1	.3	.4	.7	.8	0.0	0.0	0.0	0.0
Washington	7.6	11.2	17.8	40.0	0.0	0.0	1.1	0.0	0.0	0.0	0.0	0.0
Wyoming	3.7	5.9	5.1	8.7	.5	.7	2.0	1.8	.1	.1	.2	.2
TOTAL GROUND WATER[a]	226.1	435.7	674.8	825.2	46.7	99.1	235.3	237.7	12.3	11.4	23.4	24.6

(Continued)

Appendix Table 8.6 (Cont.)

Region/State	Electricity 1974	Electricity 1977	Electricity 1980	Electricity 1983	Diesel 1974	Diesel 1977	Diesel 1980	Diesel 1983	Gasoline 1974	Gasoline 1977	Gasoline 1980	Gasoline 1983
Surface Water:												
Arizona	0.0	0.0	0.0	0.0	0.0	0.0	0.0	0.0	0.0	0.0	0.0	0.0
California	4.4	10.0	9.1	8.0	b	.4	0.0	0.0	0.0	0.0	0.0	0.0
Colorado	b	.1	.2	.2	0.0	0.0	0.0	0.0	b	b	b	b
Idaho	7.2	7.3	14.9	15.5	.6	.8	1.9	1.7	.2	.2	.5	.6
Kansas	b	.1	.1	.1	.1	.6	1.5	1.5	0.0	0.0	0.0	.2
Montana	1.9	2.4	9.2	12.6	.5	.7	2.6	2.8	.2	.2	.8	.7
Nebraska	.2	2.4	2.7	3.5	0.0	2.9	5.4	5.3	.5	.1	.2	.2
Nevada	.1	.4	.7	.9	.1	.1	.3	.4	0.0	0.0	0.0	0.0
New Mexico	.1	.1	.5	.9	0.0	0.0	0.0	0.0	0.0	0.0	0.0	0.0
North Dakota	b	.1	.1	.2	b	b	b	b	b	0.0	0.0	0.0
Oklahoma	0.0	.3	.4	.2	.1	.7	2.2	2.4	0.0	.2	.5	.5
Oregon	5.8	6.6	24.0	37.2	0.0	0.0	0.0	0.0	0.0	0.0	0.0	0.0
South Dakota	.5	2.0	3.0	3.7	.6	1.7	3.9	4.3	.2	.2	.6	.6
Texas	2.7	3.8	4.2	6.2	.2	.3	.5	.5	.2	.2	.4	.3
Utah	2.2	3.6	2.2	3.1	.4	.5	.3	.3	0.0	0.0	0.0	0.0
Washington	9.1	13.3	45.3	47.3	0.0	0.0	0.0	0.0	0.0	0.0	0.0	0.0
Wyoming	.1	.8	.4	.8	b	.1	.1	b	b	b	b	b
TOTAL SURFACE WATER[a]	34.3	53.3	117.0	140.4	2.6	8.8	18.7	19.2	1.3	1.1	3.0	3.1
TOTAL GROUND AND SURFACE WATER[a]	260.4	489.0	791.8	965.6	49.3	107.9	254.0	256.9	13.6	12.5	26.4	27.7

(Continued)

Appendix Table 8.6 (Cont.)

Region/State	Natural Gas				Liquefied Petroleum Gas			
	1974	1977	1980	1983	1974	1977	1980	1983
	Million Dollars							
Groundwater:								
Arizona	10.8	27.4	48.5	108.4	0.0	0.0	0.0	0.0
California	.7	1.7	7.1	9.7	0.0	0.0	0.0	0.0
Colorado	2.1	2.3	5.1	9.3	.5	.7	1.0	1.3
Idaho	.1	.2	.4	.7	.1	.2	.3	.3
Kansas	14.5	19.1	60.7	101.9	4.4	6.9	15.8	19.9
Montana	0.0	0.0	0.0	0.0	[b]	.1	.1	.2
Nebraska	4.6	15.3	36.1	46.6	25.5	21.8	50.6	73.3
Nevada	0.0	0.0	0.0	0.0	.1	0.0	0.0	0.0
New Mexico	11.3	20.8	29.6	38.6	4.7	7.6	13.6	12.7
North Dakota	0.0	0.0	0.0	0.0	0.0	0.0	0.0	0.0
Oklahoma	4.4	8.9	21.7	21.7	3.1	6.6	10.0	9.7
Oregon	0.0	0.0	0.0	0.0	0.0	0.0	0.0	0.0
South Dakota	0.0	0.0	0.0	0.0	.3	.7	1.4	1.8
Texas	45.3	77.8	147.0	198.4	9.4	14.5	19.6	31.5
Utah	0.0	0.0	0.0	0.0	.5	.6	1.0	1.3
Washington	0.0	0.0	0.0	0.0	0.0	0.0	0.0	0.0
Wyoming	.1	.1	.2	.3	.1	.2	.4	.4
TOTAL GROUND WATER[a]	93.9	173.6	356.4	535.6	48.7	59.9	113.8	152.4

(Continued)

Appendix Table 8.6 (Cont.)

Region/State	Natural Gas 1974	1977	1980	1983	Liquefied Petroleum Gas 1974	1977	1980	1983
Surface Water:								
Arizona	0.0	0.0	0.0	0.0	0.0	0.0	0.0	0.0
California	0.0	0.0	0.0	0.0	0.0	0.0	0.0	0.0
Colorado	0.0	0.0	0.0	0.0	0.0	0.0	0.0	b
Idaho	0.0	0.0	0.0	.2	b	0.0	0.0	.1
Kansas	0.0	0.0	0.0	2.5	.2	.4	.9	2.6
Montana	0.0	0.0	0.0	0.0	.1	0.0	.2	.3
Nebraska	0.0	.7	1.4	1.8	1.7	1.1	2.0	2.5
Nevada	0.0	0.0	0.0	0.0	0.0	0.0	0.0	0.0
New Mexico	0.0	b	b	b	0.0	0.0	0.0	0.0
North Dakota	0.0	0.0	b	b	.1	0.0	0.0	0.0
Oklahoma	0.0	.1	.3	.5	.6	2.0	3.3	4.2
Oregon	0.0	0.0	0.0	0.0	0.0	0.0	0.0	0.0
South Dakota	0.0	0.0	0.0	0.0	.8	1.3	2.0	2.5
Texas	2.6	4.4	6.3	8.8	1.2	1.8	1.8	2.4
Utah	0.0	0.0	0.0	0.0	0.0	.8	.5	.5
Washington	0.0	0.0	0.0	0.0	0.0	0.0	0.0	0.0
Wyoming	0.0	0.0	b	b	b	b	b	b
TOTAL SURFACE WATER[a]	2.6	5.2	8.0	13.8	4.7	7.4	10.7	15.1
TOTAL GROUND AND SURFACE WATER[a]	96.5	178.8	364.4	549.4	53.4	67.3	124.5	167.5

[a]Total may not add due to rounding. [b]Less than $50,000.

Appendix Table 8.7
Quantity of on-farm pumped water in the 17 western states 1974-1983

State	1974	1977	1980	1983
	------- Million Acre Feet --------			
Arizona	5.1	5.1	5.1	5.1
California	14.1	15.2	14.8	14.8
Colorado	1.8	1.9	1.9	1.9
Idaho	5.2	5.9	6.2	6.3
Kansas	3.9	5.4	6.1	6.1
Montana	.9	1.0	1.2	1.2
Nebraska	8.3	11.0	12.1	13.1
Nevada	1.0	1.0	1.0	1.0
New Mexico	2.3	2.2	2.4	2.3
North Dakota	.1	.1	.1	.2
Oklahoma	1.3	1.5	1.6	1.4
Oregon	2.7	2.9	3.1	3.4
South Dakota	.2	.4	.5	.5
Texas	13.2	13.9	12.2	11.6
Utah	.6	.6	.4	.4
Washington	4.0	4.2	5.3	6.3
Wyoming	.4	.4	.5	.5
17 Western States	65.1	72.7	74.5	76.1

Appendix Table 8.8
Prices used for energy cost calculations, by region and state

	Electricity $ per kwh				Diesel $ per gallon				Gasoline $ per gallon			
Region/State	1974	1977	1980	1983	1974	1977	1980	1983	1974	1977	1980	1983
Arizona	.020	.021	.045	.045	.37	0.00	1.05	1.00	.48	0.00	1.17	1.19
California	.020	.042	.060	.062	.37	0.50	0.97	0.97	.48	0.00	0.00	1.21
Colorado	.023	.035	.060	.075	.36	0.45	1.01	0.95	.46	0.52	1.16	1.20
Idaho	.016	.016	.030	.030	.37	0.45	0.98	0.97	.48	0.50	1.18	1.22
Kansas	.020	.035	.055	.065	.37	0.43	0.98	0.97	.47	0.50	1.16	1.21
Montana	.016	.016	.040	.050	.35	0.40	1.02	0.98	.48	0.45	1.22	1.25
Nebraska	.020	.045	.050	.065	.34	0.45	0.99	0.96	.49	0.51	1.19	1.27
Nevada	.015	.030	.050	.062	.40	0.45	1.06	1.05	.50	0.00	0.00	1.24
New Mexico	.020	.035	.050	.076	.36	0.00	0.99	1.00	.47	0.00	1.12	1.20
North Dakota	.020	.037	.038	.065	.37	0.45	1.03	1.00	.48	0.55	1.20	1.28
Oklahoma	.023	.030	.045	.070	.35	0.45	1.02	1.01	.45	0.55	1.17	1.16
Oregon	.013	.014	.030	.042	.35	0.00	0.97	0.93	.46	0.00	0.00	1.18
South Dakota	.020	.040	.045	.050	.37	0.43	1.05	0.98	.47	0.49	1.13	1.26
Texas	.023	.030	.045	.065	.34	0.45	1.00	0.93	.43	0.55	1.14	1.12
Utah	.015	.025	.035	.045	.40	0.45	0.97	1.02	.50	0.00	1.16	1.21
Washington	.008	.011	.014	.026	.35	0.00	0.97	0.99	.49	0.00	0.00	1.27
Wyoming	.023	.035	.030	.050	.36	0.45	1.02	0.99	.48	0.50	1.19	1.18

(Continued)

Appendix Table 8.8 (Cont.)

Region/State	Natural Gas $ per mcf				Liquefied Petroleum Gas $ per gallon			
	1974	1977	1980	1983	1974	1977	1980	1983
Arizona	a	1.50	2.50	5.30	.30	0.00	0.73	0.97
California	a	0.00	4.37	5.90	.30	0.00	0.00	0.92
Colorado	a	1.15	2.50	4.20	.29	0.39	0.59	0.71
Idaho	a	0.00	2.00	4.50	.29	0.00	0.63	0.80
Kansas	a	0.80	2.50	4.20	.26	0.32	0.55	0.69
Montana	a	0.00	0.00		.30	0.35	0.60	0.73
Nebraska	a	1.20	2.50	3.00	.27	0.35	0.55	0.66
Nevada	a	0.00	0.00		.32	0.00	0.00	0.92
New Mexico	a	1.80	2.50	3.18	.28	0.41	0.69	0.77
North Dakota	a	0.00	2.50		.30	0.00	0.59	0.72
Oklahoma	a	1.20	2.50	3.50	.28	0.44	0.68	0.76
Oregon	a	0.00	0.00		.30	0.00	0.00	0.77
South Dakota	a	0.00	2.50	3.50	.29	0.41	0.60	0.72
Texas	a	1.30	2.50	3.50	.29	0.44	0.60	0.76
Utah	a	0.00	0.00		.32	0.41	0.70	0.86
Washington	a	0.00	0.00		.30	0.00	0.00	1.05
Wyoming	a	0.00	3.00	4.00	.27	0.43	0.72	0.77

[a]Not available.

Appendix Table 8.9
Estimated center pivot irrigated area in the 17 western states, 1983

State	Center Pivot Irrigated Area			
	1980		1983	
	All	Low Pressure	All	Low Pressure
	------------- 1,000 Acres -------------			
Arizona	28	5	30	5
California	0	0	0	0
Colorado	600	240	633	317
Idaho	187	10	224	47
Kansas	987	293	987	293
Montana	60	0	66	20
Nebraska	2,356	236	2,815	282
Nevada	10	0	10	3
New Mexico	225	74	246	81
North Dakota	117	12	137	55
Oklahoma	208	0	155	93
Oregon	156	8	233	23
South Dakota	226	11	239	12
Texas	570	86	543	135
Utah	11	0	11	0
Washington	389	4	410	287
Wyoming	86	13	93	23
TOTAL	6,216	992	6,832	1,677

Appendix Table 8.10
Method of water distribution by farms receiving water from irrigation organization

State	Furrows and Ditches	Flooding	Sprinkler System	Other
	------------- 1,000 Acres	----------		
Arizona	318	285		[a]
California	2,577	2,010	969	114
Colorado	1,329	907	215	3
Idaho	1,198	689	696	9
Kansas & Oklahoma	96	23	2	0
Montana	415	890	302	24
Nebraska	657	91	23	3
Nevada	70	182	5	0
New Mexico	308	147	33	[a]
North Dakota	14	18	2	0
Oregon	181	485	373	1
South Dakota	23	64	9	0
Texas	553	493	16	6
Utah	474	621	179	1
Washington	481	29	650	2
Wyoming	535	614	89	3
17 Western States	9,229	7,548	3,575	166

Source: 1978 Census of Agriculture, Irrigation, volume 4, AC78-IR.

[a]Less than 1,000 acres.

Appendix Table 8.11
Expenditures for energy by pumping irrigation organizations in the 17 western states, 1978

State	Type of Energy				
	Electricity	Natural Gas	Diesel	Gasoline	LPG
	------------- 1,000 Dollars ----------				
Arizona	5,639	262	94	132	6
California	20,656	2,487	96	388	74
Colorado	148	7	19	58	13
Idaho	4,490	[a]	88	168	[a]
Kansas	[a]	[a]	[a]	[a]	[a]
Montana	184	[a]	15	27	[a]
Nebraska	156	[a]	9	24	[a]
Nevada	23	0	0	0	0
New Mexico	78	126	[a]	15	[a]
North Dakota	18	[a]	[a]	10	[a]
Oklahoma	[a]	[a]	[a]	[a]	[a]
Oregon	952	1	40	72	1
South Dakota	114	0	[a]	[a]	0
Texas	3,547	1,643	198	166	28
Utah	546	10	15	22	0
Washington	1,635	[a]	76	236	[a]
Wyoming	92	1	12	27	[a]
17 Western States	38,278	4,537	662	1,345	123
Percent	85	10	1.5	3	.5

[a]Not available.

9
Energy Requirements for Well Installation

Richard H. Cuenca

INTRODUCTION

Many studies have quantified estimates of energy consumption for various components of agricultural production and commodity processing. These estimates of energy use included the energy consumed in irrigation (English, et al., 1982a; English, et al., 1982b). Also, the embodied energy required to manufacture materials that go into the construction of irrigation system has also been estimated (Cuenca, et al., 1981).

These studies on energy required for irrigation have generally omitted the energy consumption associated with groundwater well installation. Literature searches and discussions with groundwater consulting engineers and well drillers indicated that very little information was available to quantify energy consumption or well installation. Well drillers expressed an interest in this type of information as an aid to formulating estimates for drilling project bids. The USDA desired information on energy consumption in well drilling and installation to add to its data base of energy consumption in agricultural operations. The information discussed in this chapter was collected to satisfy those general interests.

The purpose of the study reported here was to quantify energy consumption in the installation of groundwater wells. The data collected were for irrigation wells but the results should be equally valid for other groundwater uses. Energy consumption was to be quantified for all normal phases of well installation including well drilling, well development, and well testing. Finally, an interactive computer program was developed for estimating energy requirements for well installation over a wide range of conditions, including the use of different drilling methods and drilling in different geologic formations. The remainder of this section gives a brief description of methods of well drilling, well development, and well testing.

METHODS OF WELL DRILLING

The three major methods of well drilling covered in this work are cable-tool, mud rotary, and air rotary. These are the predominant methods used for drilling irrigation wells in the United States. Cable-tool is a percussion type method, while mud and air rotary methods are similar in that drilling bits are used in combination with hydraulic fluids in both methods. (The rotary percussion method uses a combination of successive impacts delivered through a rotary drilling rig. It is not widely used for irrigation wells and will not be discussed in this report.) The methods differ in their recommended applicability with regard to type of geologic formation and well diameter. Tables 9.1 and 9.2 indicate applicability and typical sizes of wells drilled by the cable-tool and rotary methods. Following are basic descriptions of the three drilling methods examined in this study. The listed characteristics of rotary drills apply to both mud rotary and air rotary methods. More detailed description of these methods can be found in references on groundwater and wells (Johnson Division; Todd; Marino and Luthin).

Cable-Tool Method

The cable-tool drilling method is a percussion type method that relies on the impact of the drill bit to break up the formation material so that it may be removed from the bore hole. The impact to the bottom of the hole is provided by what is termed a string line of tools suspended from pulleys by steel cable which is alternately raised and lowered. A drawing of a typical drill string composed of a drill bit, drill stem, jars, and a rope socket for attachment to the drilling line is shown in Figure 9.1. A photo of an actual cable-tool drilling rig is shown in Figure 9.2.

Table 9.1
Performance of drilling methods in various types of geologic formations (taken from Speedstar Division)

Type of Formation	Drilling Method: Cable Tool	Rotary	Rotary Percussion[a]
Dune Sand	Difficult	Rapid	NR
Loose Sand and Gravel	Difficult	Rapid	NR
Quicksand	Difficult, except in thin streaks. Requires a string of drive pipe.	Rapid	NR
Loose Boulders in Alluvial Fans or Glacial Drift	Difficult; slow but generally can be handled by driving pipe.	Difficult, frequently impossible	NR
Clay and Silt	Slow	Rapid	NR
Firm Shale	Rapid	Rapid	NR
Sticky Shale	Slow	Rapid	NR
Brittle Shale	Rapid	Rapid	NR
Sandstone, Poorly Cemented	slow	Slow	NR
Sandstone, Well Cemented	Slow	Slow	NR
Chert Nodules	Rapid	Slow	NR
Limestone	Rapid	Rapid	Very rapid
Limestone with Small Cracks or Fractures	Rapid	Slow	Very rapid
Limestone, Cavernous	Rapid	Slow to impossible	Difficult
Dolomite	Rapid	Rapid	Very rapid
Basalts, Thin Layers in Sedimentary Rocks	Rapid	Slow	Very rapid

(Continued)

Table 9.1 (Cont.)

	Drilling Method		
Type of Formation	Cable Tool	Rotary	Rotary Percussion[a]
Basalts, Thick Layers	Slow	Slow	Rapid
Metamorphic Rocks	Slow	Slow	Rapid
Granite	Slow	Slow	Rapid

[a]NR: not recommended.

Water, either naturally occurring or pumped into the hole, is required to mix the loosened formation material into a slurry. After approximately three feet penetration into the formation, the broken material, termed cuttings, must be removed. The drill string is removed and a bailer or sand pump is inserted into the hole using the sand line indicated in Figure 9.1. A bailer is a pipe about 10 to 30 feet in length with a one-way valve at the bottom. When lowered in the hole, the valve opens up due to the upward pressure of fluid and the bailer fills with the slurry containing the cuttings. It is pulled to the top of the hole by the sand line and emptied. The sand pump is similar to the bailer except that it has a piston type device which when drawn upward produces a suction which draws in the slurry.

Cable-tool drilling can be used in all formations, although it is relatively slow in poorly consolidated formations, such as sand or sandstone or formations of extreme hardness, such as granite or metamorphic rocks. The method can be applied to shallow or deep holes of various diameters although is most applicable to relatively shallow holes, less than 150 feet, of moderate or larger diameter, 6 inches or more. Little water is required and the geologic formation can be accurately logged during the drilling operation. Initial equipment and operating costs

Table 9.2

Water well construction methods and applications (adapted from U.S. Soil Conservation Service, 1969)

Method	Materials for Which Best Suited	Water Table Depth for Which Best Suited (feet)	Usual Maximum Depth (feet)	Usual Diameter Range (inches)	Usual Casing Material	Customary Use	Yield, Yd^3/day [a]	Remarks
Cable Tool	Unconsolidated and consolidated medium hard and hard rock	Any depth	1,470[b]	3-24	Steel or wrought-iron pipe	All uses	20-20,000	Effective for water exploration. Requires casing in loose materials. Mud-scow and hollow rod bits developed for drilling unconsolidated fine to medium sediments.
Rotary	Silt, sand, gravel less than 2 cm; soft to hard consolidated rock	Any depth	1,470[b]	3-18	Steel or wrought-iron pipe	All uses	20-20,000	Fastest method for all except hardest rock. Casing usually not required during drilling. Effective for gravel envelope wells.

(Continued)

Table 2 (Cont.)

Method	Materials for Which Best Suited	Water Table Depth for Which Best Suited (feet)	Usual Maximum Depth (feet)	Usual Diameter Range (inches)	Usual Casing Material	Customary Use	Yield, Yd^3/day[a]	Remarks
Reverse-Circulation Rotary	Silt, sand, gravel, cobble	6–100	200	16–48	Steel or wrought-iron pipe	Irrigation, industrial, municipal	3,300–26,000	Effective for large-diameter holes in unconsolidated and partially consolidated deposits. Requires large volume of water for drilling. Effective for gravel envelope wells.
Rotary-Percussion	Silt, sand, gravel less than 5 cm; soft to hard consolidated rock	Any depth	2,000[b]	12–20	Steel or wrought-iron pipe	Irrigation, industrial, municipal	3,300–20,000	Now used in oil exploration. Very fast drilling. Combines rotary and percussion methods (air drilling) cuttings removed by air. Would be economical for deep water wells.

[a] Yield influenced primarily by geology and availability of groundwater.
[b] Greater depths reached with heavier equipment.

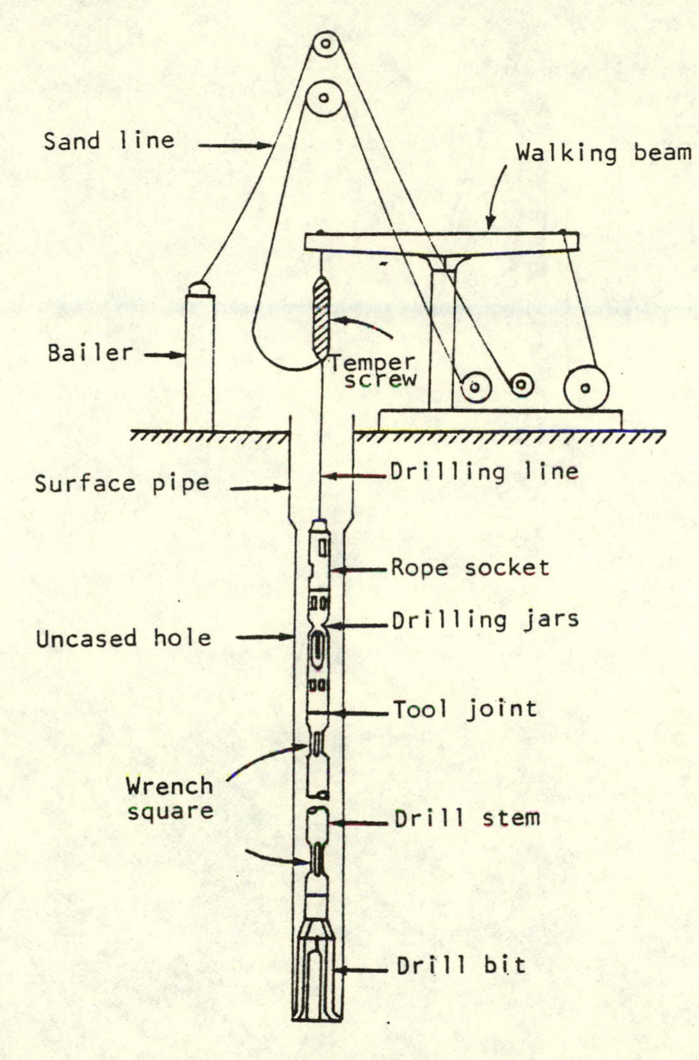

Figure 9.1 Schematic of cable-tool drilling method (Marino and Luthin)

Figure 9.2 Cable-tool drilling rig in operation. The driller holds the cable to sense the progress of the drilling.

for many types of formations are modest compared to typical rotary drilling methods.

Rotary Methods. The principles and hydraulics of mud and air rotary methods are similar, the only difference being the fluid used in the drilling operation is either a synthetic mud or air. The synthetic mud is produced by mixing bentonite with water. The consistency of the mud varies depending on the density of the cuttings to be brought to the surface. Denser particles require mud with a higher concentration of bentonite. Because of similarity of operation, both rotary drilling methods will be described in this section.

The main components for rotary drilling (see Figure 9.3) are a pump to pressurize the fluid, hollow drill pipes to transmit the pressurized fluid, and a hollow drill bit which grinds the formation and passes the drilling mud to a settling basin in which the cutting laden mud is allowed to pond so that the cuttings drop to the bottom. In normal rotary drilling, the pressurized drilling fluid is forced through the drill bit and immediately upon exiting the bit comes into contact with the cuttings which are then carried upward in the annular space between the drill pipes and hole wall. After being pumped to the surface, a fraction of cuttings are removed by draining mud off the surface of an initial settling basin and letting the mud flow into a second adjacent basin. The drilling mud is then recirculated through the pump. A schematic of the hydraulic rotary drilling method is indicated in Figure 9.3. Photographs of a mud rotary drilling rig and dual settling basin are indicated in Figures 9.4 and 9.5, respectively.

The rotary drilling method is relatively fast in many types of geologic formations as indicated in Table 9.1. Alternate layers of soft and hard material in the same hole can be handled more effectively by rotary drilling methods than the cable-tool method. It is difficult to obtain representative formation samples using this method since the material is crushed in the drilling process and cuttings from different layers are mixed in the drilling fluid. Considerable amounts of water are required for the mud rotary method compared to cable-tool, especially in gravel and fissured rock formations. As indicated in Table 9.1, mud rotary drilling in cavernous limestone is slow to impossible due to the loss of fluid down the hole. Air rotary methods, which may use a foam for the drilling fluid, have been successfully applied in formations which were not suitable for the mud rotary method.

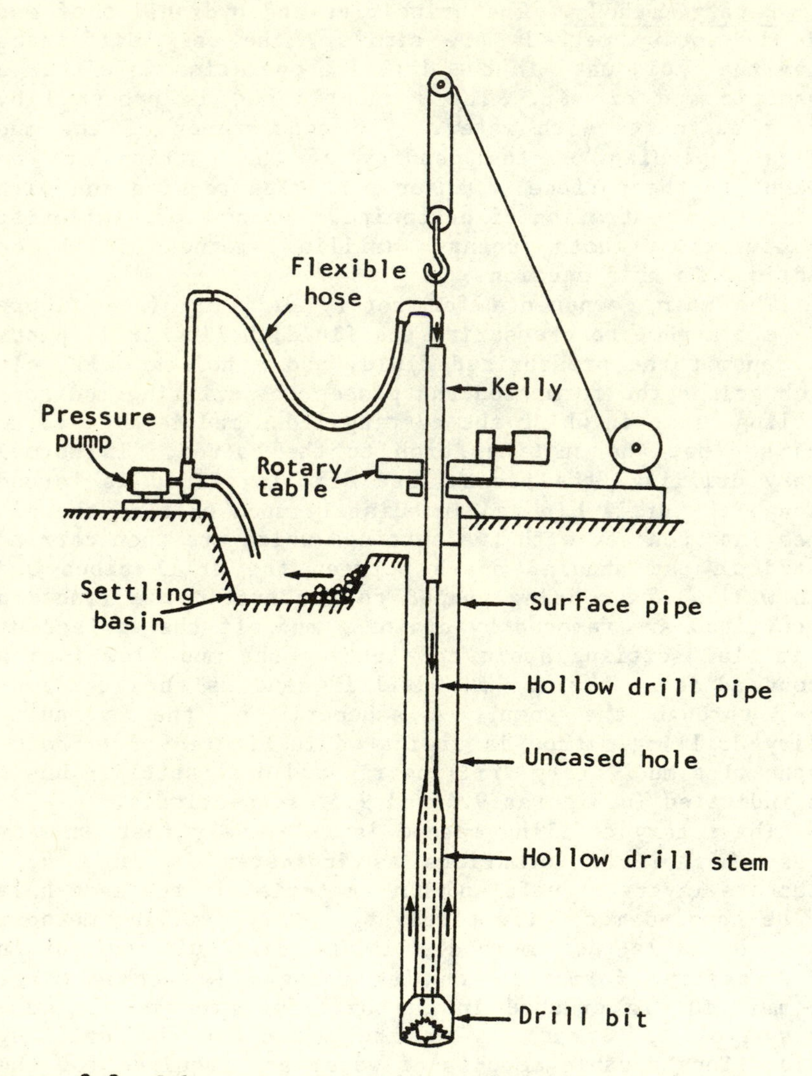

Figure 9.3 Schematic of hydraulic rotary drilling (Marino and Luthin)

Figure 9.4 Mud rotary drilling rig on location at drill site

Figure 9.5 Dual settling basin for mud rotary drilling operation

Modifications of the normal rotary procedure include reverse rotary and rotary-percussion drilling methods. The reverse rotary method operates on the same basic principles as the normal rotary method except that the drilling fluid circulates down through the annular space between the hole wall and drill pipe and is brought up the drill pipe by a suction pump. This method is shown schematically in Figure 9.6. The method is particularly well adapted to drilling large diameter wells in unconsolidated formations.

Well Development

After well drilling is complete in unconsolidated formations, perforated pipe or well screen is placed down the hole to intercept the water bearing aquifer. An operation termed well development is next performed. The purpose of this operation is to correct damage to the formation caused by clogging and compaction of the aquifer material during the drilling process. This is accomplished by removing fine particles in the aquifer adjacent to the water bearing portion of the well. Removing the fine particles, through well development, protects future pumping operations and increases the permeability of the formation in the vicinity of the well.

Fine particles and sand are extracted through the well screen or perforations and pumped out of the well so that under normal operations the well will flow sand free. There are several ways of developing a well after drilling. The three usual methods of well development are surging, airlift, and jetting. The reader interested in details about these methods is referred to Johnson Division and Speedstar Division.

Well Testing

An integral part of completing well installation is testing the well. Well tests are performed for two reasons: (1) to determine hydraulic characteristics of the aquifer material, and (2) to determine operating characteristics of the well. The most common testing methods are the specific capacity test, constant rate test, and step drawdown test.

Specific Capacity Test. This test measures variation in water level drawdown with increasing flow rate from a

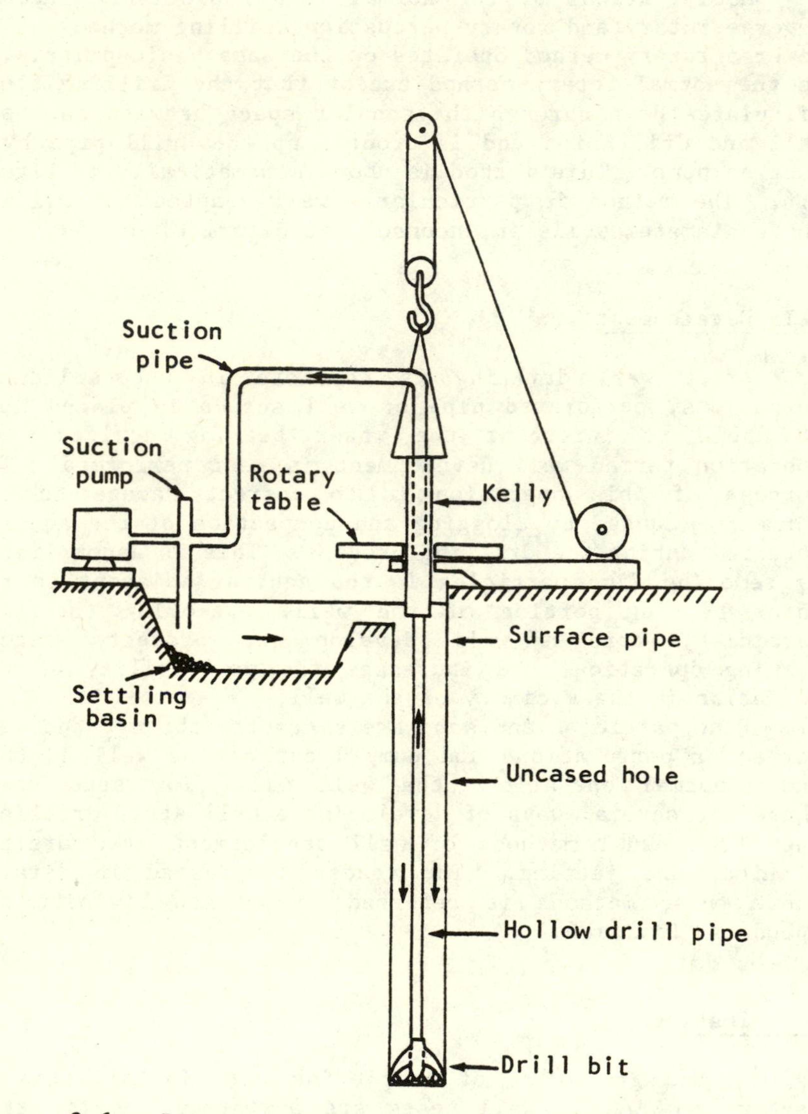

Figure 9.6 Reverse hydraulic rotary method showing circulation of water and cuttings (Marino and Luthin)

well. It gives an indication of the operating characteristics of a well. The pump is started at 20 percent of design capacity and drawdown is measured after a fixed time interval. Discharge is then incrementally increased to 40, 60, 80, 100, and 120 percent of design discharge and the drawdown for each level of discharge measured after pumping for the same fixed time increment.

The specific capacity of the well is determined from this test. The specific capacity is defined as the rate of discharge available from the well per unit depth of drawdown. The specific capacity can be useful in determining the operating characteristics of the well system. It should be noted that application of the specific capacity to estimate operating characteristics of the well system is limited to the time increments for which the specific capacity test was conducted. This is due to the fact that drawdown in a well generally increases with time for a given level of discharge. The time increment for the specific capacity test should therefore be chosen to agree with the expected operating time of the well system.

Constant Rate Test. The constant rate test is performed by measuring the variation in the level of drawdown with time for a given level of discharge. If a single constant rate test is to be run, it should be at the design capacity of the system. If multiple constant rate tests are to be run, discharge can be varied between 60 and 125 percent of design discharge. It is a good practice to run the constant rate test for at least the expected operating time of the system. The constant rate test is used to determine both operating characteristics of the well system and hydraulic properties of the aquifer. The hydraulic properties of the aquifer are used for well design and to estimate response of the well-aquifer system in hydraulic simulation models.

Step Drawdown Test. The step drawdown test involves running the pump at incremental rates of discharge and plotting the variation of drawdown with time for each level of discharge. Typical tests involve four or five levels of discharge. These levels are selected by taking the maximum expected discharge and dividing it into four or five equal increments. The test is begun at the lowest level of discharge and progresses to the maximum level. The time of testing at each level is variable and increases with increasing discharge. Normal practice is to not run the test for less than 30 minutes at the first level of discharge, 1 hour for the second step, 1.5 hours for the third step, and so on in half hour increments. Drawdown is constantly monitored and recorded during the test. After completing the recordings at the final level of discharge, the pump is turned off and the time noted. The recovery water levels, also termed residual drawdown, may be recorded as a function of time after the pump is turned off for certain types of analysis.

Analysis of the step drawdown test results are more complex than for the specific capacity or constant rate tests. The results are used to calculate the aquifer transmissivity and the relationship between drawdown and well discharge.

WELL DRILLERS' SURVEY

Data Collection

A survey of well drillers was conducted to gather data for an accurate estimate of energy consumption during the well drilling process (Cuenca, et al., 1984). The first part of the questionnaire asks for drilling rig specifications. The second section deals with drilling speed in terms of penetration rate for different geologic formations. The formations covered were sand and gravel, sandstone, limestone, fractured basalt, and basalt. The drilling speed also had to be specified for different well diameters from 6 inches to 24 inches. Drillers were asked for information on fuel requirements during drilling for diesel and gasoline fuel types. Also, the time required for well development for different soil types and well diameters was determined.

Eighteen different well drillers in Oregon and Nevada, with many years experience in domestic and irrigation well drilling, were questioned. Information was collected from 13 drillers on the cable tool method, 13 drillers on the air rotary method and 6 drillers on the mud rotary method.

Data Evaluation

The key data for quantifying energy consumption in well drilling was drilling speed versus borehole size and geologic formation. Three alternate forms of data evaluation using statistical methods were performed. All three forms of analysis are described in the Project Completion

Report (Cuenca, et al., 1984). This section will only describe the averaged data method of analysis which will remain the recommended method until a larger data base can be developed. The computer program for energy consumption has been developed so that any of the three forms of regression for drilling speed can be accessed through a subroutine.

Due to the high variability in the original data set and the resulting poor coefficient of determination for various regression analysis schemes, the mean drilling speed for each borehole diameter and geologic formation was chosen to give some direction to the analysis. The final regression equations using these data have a high coefficient of determination because they are being fit with a limited number of observations. The resulting coefficient of determination is not indicative of the functional fit through the original data, but through the mean data.

The averaged data used in the regression analysis for drilling speed as a function of borehole diameter are indicated in Table 9.3 for the three drilling methods and five geologic formations. Note that only diameters up to 12 inches were considered because of insufficient data to obtain reasonable regression relations for the large diameters. These data can be used only with extrapolation for boreholes greater than 12 inches in diameter.

A certain amount of subjective data selection was exercised in developing this data set. Values for drilling speed which were, by inspection, exceedingly higher or lower than the majority of the values reported were excluded from the averaging process. This procedure was not statistically objective but did result in a set of values which were verified as being reasonable by an experienced driller, Mr. Dutch Jungmann. Details of the reasoning applied for exclusion of particular data elements under the three drilling methods are indicated in the Project Completion Report (Cuenca, et al., 1984).

For each drilling method and each geologic formation, a stepwise regression analysis was used to determine those variables most important in the regression equation for drilling speed, y, as a function of borehole diameter, x. The terms allowed to enter the regression relationship were x, x^2, x^3, x^4 and log x. In all cases, it was found that a relationship with a constant term plus two functions of x produced a satisfactory coefficient of determination. The regression equation therefore took the following form:

Table 9.3
Averaged data for drilling speed as function of drilling method, geologic formation, and borehole diameter, ft/hr

Formation Diameter (inches)	Drilling Method		
	Cable Tool	Mud Rotary	Air Rotary
Sand and Gravel			
6	3.23	28.75	21.00
8	2.94	23.25	18.48
10	2.19	18.00	13.25
12	1.97	13.50	13.14
Sandstone			
6	6.51	52.50	66.25
8	6.35	38.17	51.88
10	5.35	33.33	29.80
12	5.31	29.33	20.54
Limestone			
6	6.60	45.83	78.93
8	6.15	30.67	64.29
10	6.90	26.67	41.67
12	5.41	23.50	34.17
Fractured Basalt			
6	2.30	28.33	37.27
8	1.49	23.33	30.34
10	1.90	19.33	19.06
12	1.73	15.67	17.71
Basalt			
6	1.55	35.00	32.14
8	1.67	25.00	24.55
10	1.45	17.50	17.38
12	1.41	14.25	16.95

$$Y = a_o + a_1 b_1 + a_2 b_2$$

where:

Y = drilling speed, ft/hr
a_o, a_1, a_2 = regression coefficients
b_1, b_2 = powers of x
x = borehole diameter, in

Table 9.4 shows the final forms of the regression equations plus the coefficient of determination (R^2) for each equation.

Data Limitations

The drillers interviewed came, in general, from a limited geographic region and, therefore, had certain limitations in experience both with respect to drilling methods and geologic formations. For example, relatively few of the drillers had experience with the mud rotary drilling method. However, one of the drillers, Dutch Jungmann, had drilled a large number of wells using various drilling methods in many countries around the world. His knowledge and experience was quite useful in helping to verify the reasonableness of the data collected.

The majority of the drillers interviewed were experienced with drilling methods, drilling depths, and geologic formations found in the western United States, particularly west of the Rocky Mountains. It is expected that expansion of the survey to a wider group of drillers would improve the statistical analysis and improve confidence in application of the results. Users of the resulting computer program have the option of using a personally derived drilling speed in the computer program thereby circumventing any of the estimating methods.

COMPUTER PROGRAM FOR ENERGY REQUIREMENTS

General Description

The results of this project are most useful in the form of a convenient computer program. The computer program is interactive to allow the user various options in setting up the well drilling situation. The objectives of

Table 9.4
Regression equations for averaged well drilling data

Drilling Method	Formation	a_o	a_1	a_2	b_1	b_2	R^2
Cable Tool	Sand and Gravel	4.901	-0.2718	1.412×10^{-5}	x	x^4	0.9563
	Sandstone	2.624	-0.6716	8.553	x	$\log x$	0.9974
	Limestone	8.003	-0.0152	-2.274×10^{-5}	x^2	x^4	0.9991
	Fractured Basalt	11.267	0.7122	-17.034	x	$\log x$	0.9617
	Basalt	1.193	-0.1454	1.758	x	$\log x$	0.9615
Mud Rotary	Sand and Gravel	46.359	-2.960	1.275×10^{-4}	x	x^4	1.0000
	Sandstone	185.453	7.878	-231.916	x	$\log x$	0.9935
	Limestone	180.686	8.507	-239.525	x	$\log x$	0.9710
	Fractured Basalt	58.374	-5.374×10^{-5}	-38.531	x^4	$\log x$	1.0000
	Basalt	71.255	-6.224	-8.521×10^{-4}	x	x^4	1.0000
Air Rotary	Sand and Gravel	36.371	-2.567	3.520×10^{-4}	x	x^4	0.9462
	Sandstone	131.995	-10.927	9.067×10^{-4}	x	x^4	0.9811
	Limestone	145.573	-11.110	1.017×10^{-3}	x	x^4	0.9819
	Fractured Basalt	72.958	-5.984	7.741×10^{-4}	x	x^4	0.9678
	Basalt	65.001	-5.632	9.335×10^{-4}	x	x^4	0.9932

the program are to quantify the results of the well drilling in terms of drilling speed and drilling time and to quantify the energy requirements in various phases of the drilling process. The energy calculations include drilling energy, installation energy including well development, energy for well testing, transportation energy, and total energy.

The program allows the user to input options related to a specific drilling situation. The following is a partial list of options available to the user:

1. type of drilling method,
2. number, depth and type of geologic formation,
3. well diameter,
4. type of well screen material,
5. installation of gravel pack,
6. method of well development,
7. type and number of support vehicles,
8. travel distance to well site, and
9. type of well test.

The program has been developed to run on a MS-DOS system using MS-BASIC (i.e., IBM compatible). Comments and sub-routines have been used extensively in the program to aid in future modification. All output data are tabulated as indicated in the sample program runs at the end of this section.

Terminal Instructions

The program to compute energy requirements for well installation is fully interactive. All input data may be set up by responding to the series of questions which appears on the terminal. The questions which are asked of the user are demonstrated specifically for use of the mud rotary drilling method. Using either of the other two drilling methods would result in a question list almost identical to the one demonstrated.

The following questions are indicated in the same form that they appear on the terminal and are basically self-explanatory. When a question or series of questions is repeated in a loop, it has been noted in the listing which follows. Notes have also been inserted to indicate the different set of questions which would be asked depending on a previous answer, e.g., whether or not a well test was of a constant rate or variable discharge.

Following are the questions and comments indicated to the user when the program is executed.

```
DO YOU WANT TO LOAD DATA FROM AN EXISTING FILE (Y/N)
```

If response is N, next question is,

```
DO YOU WANT TO INPUT DEFAULT DATA FILE (Y/N)
```

If response is N, program proceeds to ask all questions necessary to set up a new drilling situation.

```
ENTER DRILLING METHOD TO BE CONSIDERED
  1 = AIR ROTARY, 2 = MUD ROTARY, 3 = CABLE TOOL
DRILLING METHOD = ?
ENTER WELL DIAMETER (INCHES)
ENTER DEPTH OF WELL (FT)
NUMBER OF DIFFERENT TYPES OF GEOLOGIC FORMATIONS TO BE
  DRILLED
```

The following ten lines are printed for each geologic formation:

```
GEOLOGIC FORMATION NUMBER
  1 = SAND AND GRAVEL
  2 = SANDSTONE
  3 = LIMESTONE
  4 = CLAY
  5 = FRACTURED BASALT
  6 = BASALT
  7 = OTHER (DRILLING SPEED MUST BE KNOWN)
GEOLOGIC FORMATION CODE
IS CASING HAMMER TO BE USED (Y/N)
ENTER TYPE OF WELL SCREEN
  1 = WIRE SCREEN, 2 = PERFORATED PIPE, 3 = NONE
SCREEN TYPE = ?
ENTER LENGTH OF SCREEN OR PERFORATED PIPE (FT)
IS A GRAVEL PACK TO BE USED (Y/N)
WILL THE GRAVEL PACK PORTION BE UNDERREAMED (Y/N)
IS A CASING OR LINER TO BE SET (Y/N)
ENTER TYPE OF CASING OR LINER MATERIAL
  1 = PVC, 2 = STEEL
CASING OR LINER MATERIAL = ?
LENGTH OF CASING OR LINER (FT)
```

```
TIME OF WELL DEVELOPMENT (HR)
ENTER METHOD OF WELL DEVELOPMENT
  1 = BAILING (SURGING), 2 = AIRLIFT, 3 = JETTING
METHOD OF DEVELOPMENT = ?
TYPE OF AUXILIARY EQUIPMENT USED
  1 = COMPRESSOR, 2 = DESANDER, 3 = SHAKER, 4 = NONE
TYPE OF AUXILIARY EQUIPMENT = ?
IS A SUPPORT TRUCK USED (Y/N)
IS A PICKUP TRUCK USED (Y/N)
ENTER DISTANCE BETWEEN WORKSHOP AND WELLSITE (MI)
IS A WELL TEST TO BE RUN (Y/N)
TYPE OF WELL TEST
  1 = FIXED DISCHARGE, 2 = VARIABLE DISCHARGE
TYPE OF WELL TEST = ?
```

The following three lines are printed for fixed discharge text:

```
ENTER DISCHARGE (GPM)
ENTER TIME OF WELL TEST (HR)
ENTER PUMPING DEPTH (FT)
```

The following five lines are printed for variable discharge test with the last four lines printed for each discharge:

```
ENTER NUMBER OF STEPS
VALUES FOR STEP
  ENTER DISCHARGE (GPM)
  ENTER TIME OF WELL TEST (HR)
  ENTER PUMPING DEPTH (FT)
DO YOU WANT TO LOAD DATA TO A NEW FILE (Y/N)
DO YOU WANT TO RUN ANOTHER SYSTEM (Y/N)
```

Assumptions Applied

A number of assumptions were applied in the development of the computer program. Most of these assumptions were based on the experience of well drillers and taken from the survey forms. An individual well driller's experience may not agree with a particular assumption applied or may require a modification of that assumption. In addition, future developments in the field of well drilling may require revision of some of the assumptions. This section specifies the assumptions applied and indicates caution in application of these data.

Main Program. Allowable well diameters in this program are between 5 and 28 inches. However, it must be recognized that the majority of the data were collected for wells with diameters of less than 12 inches. Use of the method for wells larger than 12 inches in diameter requires extrapolation beyond the range of basic data.

Calculate Installation Energy. The energy required to weld steel casing is computed in this subroutine. This is done by first computing the time of welding. The time of welding is computed by assuming that a weld equal in length to the circumference of the casing is required for every 20 foot length of casing, the nominal length for steel casing. The speed of welding is assumed to be 0.75 inches per minute, including the entire time that the welding equipment must be running in order to complete the weld. Fuel consumption for the welder, assumed to be gasoline driven, is given as 1.5 gal/hr.

Energy equivalents of liquid fuel are used in this subroutine. The energy equivalents applied were 35.15 kwh per gallon of gasoline and 42.95 kwh per gallon of number 2 diesel fuel (Bolz and Tuve).

If the well is to have a gravel pack, but the gravel pack portion of the well is not to be underreamed, the well diameter is increased by a nominal 6.0 inches. This larger well diameter is used for all calculations, including those for drilling speed.

The casing drive speed is computed as an exponential function of well diameter. The exponential function applied was derived by fitting a regression equation through averaged responses from the drillers' survey. The time to drive the casing using a casing hammer is added to the time required to drill the well, each having the same rate of fuel consumption. Notice that the time to weld and drive the casing is based on the input casing length. If the casing is to be pulled to expose the well screen, the total length of casing used in the drilling operation should be input as the casing length.

Fuel consumption rates for various types of drilling rigs used in this subroutine are shown below:

Low Pressure Rotary Rig = 7.41 gal/hr
High Pressure Rotary Rig = 14.05 gal/hr
Cable Tool Rig = 1.44 gal/hr

For rotary drilling, the high pressure rotary rig fuel consumption rate is assumed to apply if the well is deeper than 600 feet, if a casing hammer is used, or if auxiliary equipment is used. In all other cases, low pressure rotary fuel consumption is assumed to apply. For cable tool drilling, if the well depth is greater than 600 feet, the cable-tool fuel consumption is increased to 1.80 gal/hr.

The energy required for drilling is computed by multiplying the drilling time by the drilling rig fuel consumption rate. This product is then multiplied by the energy conversion rate for the fuel. Rotary drilling rigs are assumed to use number 2 diesel fuel and cable-tool rigs are assumed to use gasoline. If a steel casing is used in the well, the drilling rig is assumed to continue to run during the required time of welding. The additional fuel consumed by the drilling rig during the time of welding is added to the drilling energy.

Calculate Drilling Speed. Of the six geologic formations listed in setting up the input file, no data were collected for the drilling speed in clay. If a geologic formation is made up of clay, the drilling speed through the clay is assumed to be the same as the drilling speed through limestone. Note that this same assumption is made for clay geologic formations regardless of the method used to compute the drilling speed.

Calculate Energy for Well Development. This subroutine employs an appropriate consumption rate for the well development method used. The cable-tool rate of 1.44 gal/hr is assumed for development by bailing. The low pressure rotary rig rate of 7.41 gal/hr is assumed for development by airlift. If the well depth is greater than 600 feet or if the development is by jetting, the high pressure rotary rate of 14.05 gal/hr is assumed. The rotary rates are always converted to the energy equivalent of number 2 diesel fuel. The cable-tool rate is converted using the fuel conversation rate for gasoline.

Calculate Well Testing Energy. Pumping plant efficiency, including pump and motor efficiency, is assumed to be 65 percent for well testing. The total dynamic head for the well test is estimated to be equal to the pumping depth during the test plus 10 feet for friction and miscellaneous head loss. This 10 feet of head loss probably under estimates friction losses in deep wells or at particularly high values of discharge.

Calculate Transportation Energy. This subroutine estimates fuel consumption in transportation vehicles. On the road, fuel consumption for rotary drilling rigs is assumed equal to 4.11 miles per gallon. For cable-tool rigs, this value is equal to 5.12 miles per gallon. The drilling rig is assumed to make one round trip between the workshop and the drilling site. The fuel conversion factor used for the drilling rig on the road assumes a gasoline engine. The support truck is also assumed to make one round trip between the workshop and the drill site. On the road fuel consumption for the support truck is assumed to be 5.69 miles per gal or using number 2 diesel fuel.

The pickup truck is assumed to make one round trip between the workshop and well site for every 8 hours of drilling time. On the road fuel consumption for the pickup truck is assumed at a rate of 9.94 miles per gallon using a gasoline engine.

Sample Program Runs

Three sample program runs are provided for illustration. The runs have been set up to demonstrate the use of the program under various operating conditions. The same general well drilling conditions are set up for each of the runs. These include geologic formations of sandstone, fractured basalt, and sand and gravel with varying depths for each formation. The well is to be 8 inches in diameter and have a gravel pack with no underreaming for the gravel pack. A steel casing is to be driven into the well using a casing hammer for the total well depth and then pulled back to expose a steel well screen for the complete length of the sand and gravel formation. Well development is to be done by jetting for a period of 6 hours. A pickup and support truck are used with 20 miles distance between the workshop and well site. A variable discharge well test is to be run with discharges of 400, 800, and 1,200 GPM. The test is to be run for 12 hours at each discharge rate and the testing depth is assumed to be 500 feet.

This test case was run for each of the three drilling methods. For each case, the averaged data regression equation was used to approximate the drilling speed. This method of approximating the drilling speed is not necessarily better than the other two methods available in the program, but it does result in drilling speeds which have been judged to be reasonable by well drillers.

The output from the three sample runs are shown in Tables 9.5, 9.6, and 9.7. Note that the output table is divided into three sections. The first section indicates the input data used to set up the drilling situation. The

Table 9.5
Energy requirements for the air rotary drilling method

Input Data

Drilling Method = Air Rotary
Diameter (in) = 8
Total Depth (ft) = 525
Well Screen = Wire
Screen Length = 75
Length of Casing or Liner (ft) = 525
Type of Casing or Liner Material = Steel
Time of Well Development = Jetting
Distance Between Workshop and Wellsite (mi) = 20
Type of Well Test = Variable Discharge

Step Number	Discharge (gpm)	Duration (hr)	Depth (ft)
1	400	12	500
2	800	12	500
3	1200	12	500

Drilling Summary

Drilling Speed Method = Averaged Data Regression Equation

Formation Type	Formation Depth (ft)	Drilling Speed (ft/hr)	Drilling Time (hr)	Estimated Drill Speed
Sandstone	400	13.85	28.88	no
Fractured Basalt	50	18.92	2.64	no
Sand and Gravel	75	13.96	5.37	no

(Continued)

Table 9.5 (Cont.)

Energy Summary

Drilling Energy (kwh) = 21,970.58
Welding Energy (kwh) = 736.1793
Energy for Well Development (kwh) = 3,620.685
Total Installation Energy (kwh) = 26,327.44
Type of Well Test = Variable Discharge

Step Number	Discharge (gpm)	Duration (hr)	Depth (ft)	Energy (kwh)
1	400	12	500	709
2	800	12	500	1,418
3	1,200	12	500	2,128

Total Test Energy (kwh) = 4,255.184
Transportation Energy (kwh) = 973.7597
Total Energy Required (kwh) = 31,556.38

second section is a summary of the drilling operation including the drilling speed and drilling time required for each geologic formation. This section shows the estimated energy required for drilling. The third section indicates the energy required for all other aspects of the operation, including welding, development, testing, transportation, and installation. This section also provides a sum of all of the individual energy components.

CONCLUSIONS AND RECOMMENDATIONS

The chapter develops a method to quantify the energy required to install groundwater wells. The specific focus of this work was towards wells used for irrigation, but the procedure developed is applicable to wells for any purpose. In addition to being useful to researchers, this information may also be applied by well drillers as an aid in the quantification of costs for bidding on projects.

Table 9.6
Example of energy requirements for the mud rotary drilling method

Input Data

Drilling Method = Mud Rotary
Diameter (in) = 8
Total Depth (ft) = 525
Well Screen = Wire
Screen Length = 75
Length of Casing or Liner (ft) = 525
Type of Casing or Liner Material = Steel
Time of Well Development = Jetting
Distance Between Workshop and Wellsite (mi) = 20
Type of Well Test = Variable Discharge

Step Number	Discharge (gpm)	Duration (hr)	Depth (ft)
1	400	12	500
2	800	12	500
3	1200	12	500

Drilling Summary

Drilling Speed Method = Averaged Data Regression Equation

Formation Type	Formation Depth (ft)	Drilling Speed (ft/hr)	Drilling Time (hr)	Estimated Drill Speed
Sandstone	400	29.94	13.36	no
Fractured Basalt	50	12.15	4.12	no
Sand and Gravel	75	9.82	7.64	no

(Continued)

Table 9.6 (Cont.)

Energy Summary

Drilling Energy (kwh) = 23,337.7
Welding Energy (kwh) = 736.1793
Energy for Well Development (kwh) = 3,620.685
Total Installation Energy (kwh) = 27,694.57
Type of Well Test = Variable Discharge

Step Number	Discharge (gpm)	Duration (hr)	Depth (ft)	Energy (kwh)
1	400	12	500	709
2	800	12	500	1,418
3	1,200	12	500	2,128

Total Test Energy (kwh) = 4,255.184
Transportation Energy (kwh) = 1,112.83
Total Energy Required (kwh) = 33,062.58

It is clear at this stage, after recognizing the variance of reported drilling speed, that further differentiation should be made as to type of drilling rig and that a broader cross-section of drillers should be surveyed. This recommendation does not affect the usefulness of the current computer program since an estimated drilling speed can be supplied by the program user. But the program would be more useful for those without drilling experience if estimates of drilling speeds could be provided in the program with greater confidence.

The computer program developed in this project is a very flexible tool for determining energy requirements in well installation. It suffers moderately from the lack of a good statistical fit on drilling speed but an estimated drilling speed can be input by an experienced driller. The program is flexible in terms of the types of well drilling and well testing situations which may be analyzed. Output from the program clearly indicates the specifics of the well drilling situation being analyzed, details regarding

Table 9.7
Energy requirements for the cable tool drilling method

Input Data

Drilling Method = Cable Tool
Diameter (in) = 8
Total Depth (ft) = 525
Well Screen = Wire
Screen Length = 75
Length of Casing or Liner (ft) = 525
Type of Casing or Liner Material = Steel
Time of Well Development = Jetting
Distance Between Workshop and Wellsite (mi) = 20
Type of Well Test = Variable Discharge

Step Number	Discharge (gpm)	Duration (hr)	Depth (ft)
1	400	12	500
2	800	12	500
3	1200	12	500

Drilling Summary

Drilling Speed Method = Averaged Data Regression Equation

Formation Type	Formation Depth (ft)	Drilling Speed (ft/hr)	Drilling Time (hr)	Estimated Drill Speed
Sandstone	400	4.78	83.67	no
Fractured Basalt	50	1.71	29.16	no
Sand and Gravel	75	1.55	48.39	no

(Continued)

Table 9.7 (Cont.)

Energy Summary

Drilling Energy (kwh) = 47,925.93
Welding Energy (kwh) = 736.1793
Energy for Well Development (kwh) = 3,620.685
Total Installation Energy (kwh) = 52,282.8
Type of Well Test = Variable Discharge

Step Number	Discharge (gpm)	Duration (hr)	Depth (ft)	Energy (kwh)
1	400	12	500	709
2	800	12	500	1,418
3	1,200	12	500	2,128

Total Test Energy (kwh) = 4,255.184
Transportation Energy (kwh) = 1,890.123
Total Energy Required (kwh) = 58,428.1

the drilling operation itself, and information about total energy consumption. Assumptions used in the development of the program can be modified by a user by editing the program. Copies of the computer program can be obtained from the author: Richard H. Cuenca, Department of Agricultural Engineering, Oregon State University, Corvallis, OR, 97331.

REFERENCES

Bolz, R.E. and G.L. Tuve (eds). "Handbook of Tables for Applied Engineering Science, 2nd Edition." CRC Press, Cleveland, Ohio, 1977.

Cuenca, R.H., G.H. Daskalakis, and J.A. Bondurant. "Energy Use Coefficients for Irrigation System Materials." Project Completion Report, USDA- SEA/AR: Cooperative Agreement No. 48- 9AHZ-9-417, June 1981.

Cuenca, R.H., A. Lehr, and J.A. Bondurant. "Energy Consumption In Well Installation." Final Report, Cooperative Agreement No. CA-59-9AHZ-1568, 1984.

English, M.E., R.H. Cuenca, K.L. Chen, R.B. Wensink, and J.W. Wolfe. "Users Manual for Computer Model to Predict Total Energy Requirements of Irrigation Systems." Project Completion Report, Office of Water Resources and Technology: Grant No. 14-34-001-6224, June 1982a.

English, M.E., R.H. Cuenca, K.L. Chen, R.B. Wensink, and J.W. Wolfe. "Analysis of Energy Used by Irrigation Systems." Agr. Eng. Dept., Oregon State Univ., Corvallis, June 1982b.

Haan, C.T. Statistical Methods In Hydrology. Iowa State University Press, Ames, Iowa, 1977.

Johnson Division, Universal Oil Products Co. Ground Water and Wells. Saint Paul, Minnesota, 1972.

Marino, M.A. and J.N. Luthin. Seepage and Groundwater. Elsevier, New York, 1982.

Speedstar Division. Well Drilling Manual. Koehring Co., Enid, Oklahoma, (no date).

Todd, D.K. Groundwater Hydrology. John Wiley and Sons, New York, 1980.

U.S. Soil Conservation Service. Engineering Field Manual for Conservation Practices. Washington, D.C., 1969.

10
Alternative Sources of Energy for Pumping

Ronald D. Lacewell, Dennis L. Larson, Ronald C. Griffin, Glenn S. Collins, and Wayne A. LePori

INTRODUCTION

Increasing costs of irrigation associated with rising real costs of traditional fuels such as electricity, natural gas and diesel have stimulated interest and research in alternative energy sources. This chapter emphasizes solar, wind, biomass, plant oils, and ethanol as alternative energy sources that could be used in irrigation pumping. However, much of the research reported here has a broader application than irrigation pumping.

SOLAR

Irrigated agricultural areas with high energy cost and few cloudy days during irrigation periods can utilize solar energy to power irrigation pumps. However, present solar power systems require further development to improve performance and reduce capital, operating, and maintenance costs before solar powered pumping will become economic in most U.S. applications (Heid and Trotter).

The cost of solar power depends on solar energy availability and collection and conversion efficiencies as well as capital, operating and maintenance costs. Collection of concentrated solar energy can improve conversion efficiencies for thermal engines and photovoltaic cells. Thus, a number of different types of solar power systems are being developed and evaluated as potential sources of power for pumping.

Agricultural Pumping Experiments

The technical and economic feasibility of using solar energy to drive irrigation pumps was examined in two application specific experiments conducted by the Universities of Arizona (Larson, 1983) and Nebraska (Sullivan and Fischbach; Sullivan, et al.). In Arizona, a 200 kw solar thermal-electric power plant constructed by Acurex Corp. with U.S. Department of Energy support was operated from 1979 to 1982. The plant included 2,140 m^2 of single axis tracking, parabolic trough collectors, thermal energy storage capacity for five hours of power conversion subsystem operation, a 200 kw organic Rankine cycle engine and synchronous generator interconnected with the utility grid system. Plant operation characterized energy performance, determined equipment shortcomings, and quantified operating and maintenance requirements.

Solar to electrical energy conversion efficiencies ranged from 4.3 percent on a sunny June day to 3.5 percent on an equinox day to 0.2 percent in December, an expected drop for a north-south oriented collector field. Net electrical energy production was 860 kwh on a sunny June day, 630 kwh on March and September days, and only 30 kwh on a sunny December day.

This solar plant operated automatically. However, various operating and maintenance activities required about 20 hours of technician effort per week. These included daily inspection, quarterly collector washing and malfunction troubleshooting efforts. Operational supplies, repair parts and maintenance services cost about $50 per year.

Design and construction of the Coolidge (Arizona) plant cost about 5.5 million dollars in 1978-1979. In 1983, a similar plant would have cost an estimated 1.5 million dollars due to design changes and improved manufacturing methods. The Coolidge plant had insufficient collector area to fully utilize the energy storage and power conversion equipment. It produced 180 mwh of electricity in 1981-1982. Increasing the collector field size and energy storage capability to permit all day operation on a sunny June day would increase the estimated solar plant cost to three million dollars. However, the larger plant would produce an estimated 550 mwh of electrical energy per year and have only slightly greater operational and maintenance requirements than the Coolidge plant.

Preliminary analysis (Larson, 1982) estimated the cost of electricity from a 550 mwh solar plant would be $.24 to

$.38 per kwh in 1983 as shown in Table 10.1. If plant cost were halved to 1.5 million dollars, the electricity would cost an estimated $.13 to $.21 per kwh. The costs vary considerably with estimates for discount and inflation rates, energy tax credits, equipment lifetime, and operating and maintenance costs.

Beyond the economic implications, operation of the Coolidge plant provided insight into management needs. For example, energy storage system cost and energy losses make thermal energy storage a costly load management method. Thus, utility interconnection can provide cost effective, trouble free load management. This means a solar electricity generation plant is best utilized if tied to a commercial power grid (electricity utility system).

Operation of the solar system showed that solar thermal-electric power plants require substantial and skilled technician attention, limiting on-farm siting to special situations. This indicates the solar thermal-electric power plants are high technology and do not function without highly trained technicians. On-farm applications are, thus, expected to be minimal. With these findings the conclusion is that energy from solar plants will not be cost competitive in the U.S.A. until capital costs and operating and maintenance requirements decrease, energy collection and conversion efficiencies increase, and/or alternative energy source prices increase substantially.

The University of Nebraska operated a 25 kw solar photovoltaic (PV) system from 1977 to 1982. The system included 514 m^2 of fixed position, non-concentrating PV cells, 90 kwh of battery storage capacity and 3 phase inverter, and power conditioning equipment to drive ac electrical motors and other equipment. Solar energy was directly converted to dc electrical energy by PV (solar) cells. System operation quantified PV system energy performance, determined operating and maintenance requirements and evaluated interconnection with irrigation pump, nitrogen fertilizer generator, and grain dryer loads.

The study found PV cells to be reliable and require little maintenance, but battery maintenance was a weekly task and inverters also required repair on occasion. Inversion from dc to ac yielded about a 10 percent loss in energy. However dc motors were found to be somewhat more expensive and require more maintenance than ac motors. Furthermore, ac motors permitted switching from the solar to a utility power system.

Table 10.1
Estimated 1983 electrical energy cost from an on-farm solar thermal-electric powerplant[a]

Plant Cost $1,000,000	Discount Rate & Rates of Energy & O&M Cost Increases[b]	Initial Energy Cost $/kwh
3	$i_e = i_m = i_d = k$[c]	.38
3	$i_e = 15\%$, $i_m = i_d = 10\%$	.24
1.5	$i_e = i_m = i_d = k$	.21
1.5	$i_e = 15\%$, $i_m = i_d = 10\%$	.13

[a] Present worth of electricity = present worth of capital + O&M cost, all electrical energy has same value, net electrical energy production = 450 mwh/yr, initial operating and maintenance costs $20,000/yr, equipment lifetime = 20 years.
[b] i_e is rate of energy cost increase, i_m is rate of operation and maintenance cost increase and i_d is discount rate.
[c] k is any percentage change.

Photovoltaic cell power systems are too costly to be competitive except in remote or special applications. In 1977, PV cell cost was $22.50 per peak watt of capacity. In 1983, cells cost about $8 per peak watt. To be competitive with electrical energy from conventional sources, the cost of PV cells must drop to $.50 to $2.00 per peak watt.

Other Solar Power Systems

Other concepts for solar thermal energy collection being developed and evaluated include solar ponds, flat plate, evacuated tube and parabolic dish collectors, and central receiver systems. Stratified, non-convecting salt ponds have been used successfully for power generation in Israeli tests. A number of simple thermal engines have been developed for use with flat plate collectors in low head, low volume pumping applications in developing countries and for such U.S. applications as stock watering (Sir

William Halcrow and Partners). These systems operate rather inefficiently, but can be cost effective if collectors are sufficiently inexpensive. Evacuated tube collectors are more expensive, but can effectively collect higher temperature heat.

Single-axis tracking collectors of the kind used at Coolidge can collect thermal energy at up to about 300°C and increase the length of the daily energy collection period. Both of these traits increase potential thermal engine output. Two-axis tracking, point focusing collectors, such as parabolic dish collectors and heliostat reflectors used in central receiver systems, can collect thermal energy at 500-600°C or higher and further increase the daily energy collection period. Parabolic dish collectors could be operated as modules, each with its own engine, or as an interconnected group. All collector types except the central receiver type require large amounts of piping to gather energy collected by a field of collectors. A central receiver system utilizes optical concentration to collect energy at a single receiver. This minimizes piping and thus piping heat losses. In addition to collector development, organic Rankine, steam Rankine, Brayton, and Stirling cycle engines are being developed for compatibility with different size applications and thermal energy collection temperatures (McNelis; Pham and Jaffe).

It is not yet clear which type solar system is most applicable to the solar powered pumping application. It appears that non-tracking systems may be most appropriate for low power (perhaps up to 10 kw) applications while central receiver systems may be most economical for larger applications (over 600 kw). In larger systems, higher energy efficiency compensates for greater system complexity to reduce energy costs.

Lower cost photovoltaic system power is being sought through development of both less expensive, but less efficient cells and more efficient, but more costly cells. Some cells operate more efficiently to produce more power with concentrated sunlight. Systems which utilize parabolic trough reflectors and Fresnel lens concentration are being tested for power generation at various U.S. sites. Concentrating systems must track the sun, and thus increase the daily energy collection period. Concentrating systems also create heat so, as with thermal power systems, yield both heat and electricity.

Energy Management

Full utilization of solar power plant output minimizes the cost of the produced energy. A stand-alone solar powered pumping plant probably can meet peak irrigation needs only with excess capacity and/or energy or water storage (Williams, Larson, and McAniff). Other on-farm energy requirements such as grain drying and livestock shelter heating can increase plant utilization. Full utilization of solar plant output might be accomplished by selling electricity to another enterprise, for example for fuel or fertilizer manufacturing.

Water storage provides one potential means for load management (Larson, 1979). Modifying irrigation demand schedules to best correspond with solar energy supply schedules can minimize reservoir requirements. Water losses from storage and reservoir, capital, maintenance, and energy costs determine the economic practicality of water storage.

All energy storage methods for load management reduce energy use efficiency and increase energy production costs. Utility interconnection effectively provides energy storage through interchange of supply and demand periods. Where supply and demand of energy can be made to correspond with utility on-peak and off-peak demand periods respectively, the interconnection can be mutually beneficial. In isolated areas, back-up biomass fired boilers or fossil fuel powered electrical generators might be cost competitive with energy storage for meeting demands when solar energy is unavailable.

The Coolidge solar power plant included thermal energy storage sufficient for five hours of power conversion system (PCS) operation. Storage tests found less than 70 percent of the thermal energy put into storage could be recovered for use by the PCS. Thus, Coolidge energy storage was used primarily to buffer differences between energy collection rate and energy demand of the PCS, which was operated at its most efficient rate capacity. Storage also permitted PCS repairs during collector system operation. All Coolidge plant electrical output was input to the utility grid; its value was not differentiated by time of day. Storage inefficiency and uniform electrical rates thus made direct coupling of solar energy collection and power generation systems most cost effective at Coolidge.

The Nebraska photovoltaic (PV) system had 90 kwh of battery storage capacity and was directly interconnected with only irrigation pumping and grain drying loads. Some energy was dumped during periods of insufficient demand. An experimental nitrogen fertilizer generator was added to more fully utilize the PV system output. However, nitrogen generator operation was found to be uneconomical.

Load management is crucial to economical use of solar powered irrigation systems. Utility interconnection is the logical solution where available. In isolated locations, changes in energy demand and the use of energy storage and/or backup energy supplies are possible ways to meet demand and minimize energy costs.

Future

Further developments to reduce cost and improve performance of solar collectors and thermal engines and/or PV cells are required for solar power to be cost competitive with other sources of power for pumping. Currently, solar power would be applicable only where other energy sources are not readily available. For example, PV systems are very modular and require relatively little maintenance. They can be located near the application and likely in a remote area not serviced by electricity, natural gas, or other commercial forms of energy. Solar thermal power plants (with the exception of very small, perhaps less than 10 kw size plants) utilize relatively complicated equipment and are expected to be operated only by large enterprises or utilities having qualified technicians. However, solar thermal power plants are not currently economically competitive and commercial ventures are not expected until some cost advantages are developed.

Energy storage is costly and energy inefficient. Therefore, energy storage from solar power plants will be used only for short-term buffering of differences between energy supply and demand. When and if solar power becomes economically competitive, it is expected that utility interconnection will be used for load management. This, of course, requires institutional agreements.

Projecting technical advances in solar power plants which will improve the cost of electrical generation, it is expected that solar energy will gradually begin to meet irrigation pumping needs, perhaps supplying 5 percent of the energy for pumping in the U.S. in the year 2000. This will occur first in the Southwest (West Texas, New Mexico, Arizona and California) due to available power per unit area from the sun. Further, areas of pumping most remote to alternative energy sources or where conventional energy is most expensive will be the first to adopt this new technology.

WIND

The discussion in this section is based primarily on a study by Hardin and Lacewell. The use of wind energy is being examined for a number of agricultural applications, including water heating (Gunkel, et al.) and for cooling and refrigeration (O'Brien, et al.). The application of importance to irrigation farmers is that of wind-assisted irrigation pumping. Large scale wind systems have been developed which are capable of providing supplemental energy to an existing electrical pumping plant. The electric motor is sized to operate the pump on a stand-alone basis. However, with sufficient wind velocity, the wind system operates to produce electricity and reduces the power drawn from the electricity utility (Clark and Schneider). The wind-assist concept appears to be a particularly attractive alternative in regions with relatively high sustained wind speeds such as the Texas High Plains and much of the Great Plains.

System Description and Overview

A wind energy project for irrigation pumping was started at the U.S. Department of Agriculture Southwestern Great Plains Research Center in Bushland, Texas, in late 1976 (Clark and Schneider). A vertical-axis wind turbine was installed which was designed to produce 40 kilowatts (kw) in a 32 mph wind and furnish power only when the wind speed is above 13 mph. A test of the system was reported for the period between 9:00 a.m. and 4:00 p.m. on September 18, 1978. The pump produced 458 gallons per minute with a total dynamic head of 344 feet. In the seven hour period reported, the wind turbine produced an average of 35.9 kw. Over the seven hour period, 65 percent of the energy required to pump water was produced by the wind, despite the fact that the long-term average wind speed for September is one of the lowest in the year (U.S. Department of Commerce).

A 1980 study shows the Texas High Plains to have considerable potential for using wind power for irrigation (Lansford, et al.). Average monthly wind power for the High Plains was estimated from historical data and ranged from 12.1 mph in August to 15.6 mph in March. Average monthly wind power along with load characteristics, were used to estimate break-even investment values for three energy price scenarios and two discount rates (7 and 10 percent). The electricity price scenarios were: (1) 20 percent increase for two years and 4 percent thereafter, (2) 8 percent for two years and 4 percent thereafter, and (3) 4 percent for two years and 2 percent thereafter. Wind-assist electric systems with sale of surplus electricity at 60 percent of electricity cost from a utility were shown to be economically viable in the high energy price scenario. At lower energy prices, wind produced energy was not competitive for prototype units. However, irrigators in a position to take advantage of tax savings could possibly justify the investment, even at lower energy prices. For example, with federal solar tax credits available, 40 percent of the first $10,000 invested by individuals is eligible as well as 25 percent for businesses with no limit (Alternative Energy Institute). Some states such as California also have energy tax credits.

The initial cost of a wind system is subject to some uncertainty as different wind machine producers are in different stages of technological development. Recent estimates of installed costs for small systems (less than 100 kw rated output) have been as low as $500 to $700 per installed kw (Gipe) and ranging up to $2,000 per kw (Alternative Energy Institute). This is due to the immaturity of the technology. As more companies move past the prototype stage into production, prices should decrease and stabilize. Also, the cost per kw is typically less for larger size units. For 25 kw units and larger, cost per kw are approaching $1,000 for several distributors (Stoiaken).

Irrigated Farm Evaluation

A detailed study of the value of wind assisted pumping on the Texas High Plains was conducted by Hardin and Lacewell. Two wind machines were analyzed, with rated outputs of 40 and 60 kilowatts. Break-even investment was estimated at discount rates of 3, 5, and 10 percent.

Purchase price of electricity by the irrigation farmer of $.05, $.075, and $.10 per kwh were evaluated, with any surplus electricity generated by the wind system sold to a power grid for $.03, $.045, and $.06 kwh, respectively. Table 10.2 shows estimated wind generated electricity that could be sold and that which could be used for irrigation pumping. The relationship between electricity sold and that used for pumping is not sensitive to depth of the groundwater table (static water level), well yield, and cropping pattern. However, the percent of total electricity required for pumping that can be satisfied by a wind system is sensitive to these factors. Also, the estimated break-even investment or maximum that could be paid for the two wind systems was not very sensitive to these same parameters. For the 40 kw machine, about 62,000 kwh of electricity was generated per year. Of this approximately 15 percent was used for pumping irrigation water which satisfied about 20 percent of the irrigation pumping energy

Table 10.2
Estimated distribution of total wind-generated electricity sold and used for irrigation

	Wind Generated Electricity			
Wind Machine	Sold	Used for Pumping	Irrigation Requirements[a]	Irrigation Utilization[b]
	1,000 kwh		---- percent -----	
40 kw Machine	52.5	9.6	20.4	14.1
60 kw Machine	78.4	13.0	27.6	12.7

Source: Hardin and Lacewell.

[a] The average percentage of total irrigation energy requirements fulfilled by wind power.
[b] The average percentage of total wind generated electricity used for irrigation.

requirements given a 200 foot lift. For a pumping lift of 100 feet, the wind system can satisfy 42 percent of total energy requirements for pumping. For the 60 kw system, over 100,000 kwh was generated per year with 13,000 kwh used for pumping (13 percent) meeting 27.6 percent of the total pumping energy needs, again assuming a 200 foot pumping lift.

An optimal (profit maximizing) irrigation schedule was determined by a linear programming model to estimate the amount of wind energy that could be used for pumping. This optimal irrigation schedule was then used as input to a simulation model that matched stochastically generated estimates of wind power available (based on historical experience) with irrigation fuel requirements by three hour time periods throughout a year. The results indicate that there are many periods of wind and no irrigation need as well as irrigation need and no wind.

When applying the wind system to an irrigated farm, the reduction in costs of energy for pumping is estimated. This reduction in pumping costs and the value of electricity sales to a power grid provides the basis for establishing break-even investment in a wind energy system. Break-even investment is the investment at which there is no economic advantage or disadvantage attached to the system. If the required investment is greater than the break-even investment, then it is not a profitable venture.

Table 10.3 shows an estimated break-even investment in a wind system for the High Plains in an irrigation assist mode. This break-even investment is the value of reduced costs for pumping as well as returns to electricity sold. The break-even investment is the maximum a farmer could pay for a wind system and not reduce his net returns. Thus, if a wind system can be purchased for less than the break-even investment, the net return position of the farmer is improved. As the purchase price of electricity goes from $.05 to $.10 per kwh, the break-even investment approximately doubles. Considering an electricity cost to the irrigation farmer of $.075 per kwh, the 40 kw machine has a break-even investment ranging from $42,600 for a 3 percent discount rate to $25,800 at a 10 percent discount rate. This is compared to $62,200 at a 3 percent discount rate for the 60 kw machine and $37,700 at a 10 percent discount rate. Thus, the break-even investment in a wind energy system is extremely sensitive to the price of electricity and to the discount rate. Using a 5 percent discount rate and $.075 per kwh electricity price, the break-even

Table 10.3
Break-even investment for a wind-assisted irrigation pumping system

Item	Price of Electricity[a]		
	$.05	$.075	$.10
	------------	$1,000	----------
40 kw Machine			
3% Discount Rate	28.3	42.6	57.2
5% Discount Rate	24.2	36.5	49.0
10% Discount Rate	17.2	25.8	34.7
60 kw Machine			
3% Discount Rate	41.6	62.2	83.1
5% Discount Rate	35.6	53.2	71.1
10% Discount Rate	25.2	37.7	50.3

[a]Purchase price for the irrigation farmer. A credit per kwh for surplus electricity sales was assumed of $.03 for a $.05 price, $.045 for a $.075 price and $.06 for a $.10 price.

investment is $36,500 for a 40 kw machine and $53,200 for a 60 kw machine. In a previous section it was noted that current capital costs of wind machines is approximately $1,000 per kw of capacity. So only at the lowest discount rates and upper levels of electricity cost would it pay for the farmer to invest in wind machines.

Figure 10.1 indicates average wind speeds across the United States. Based on economic analysis, the implication is that an average wind speed of 12 mph or greater is required for a wind energy system to be profitable (Clark). The economic analysis was based on a 1983 wind system cost of about $1,000 per kw and average price of energy sold to the electric utility of 4.5 cents per kwh. Thus, the areas of the U.S. where wind energy is a viable alternative include primarily the Great Plains, the region of most groundwater pumping in the U.S. Approximately one-third of the U.S. has sufficient wind to economically consider wind energy systems.

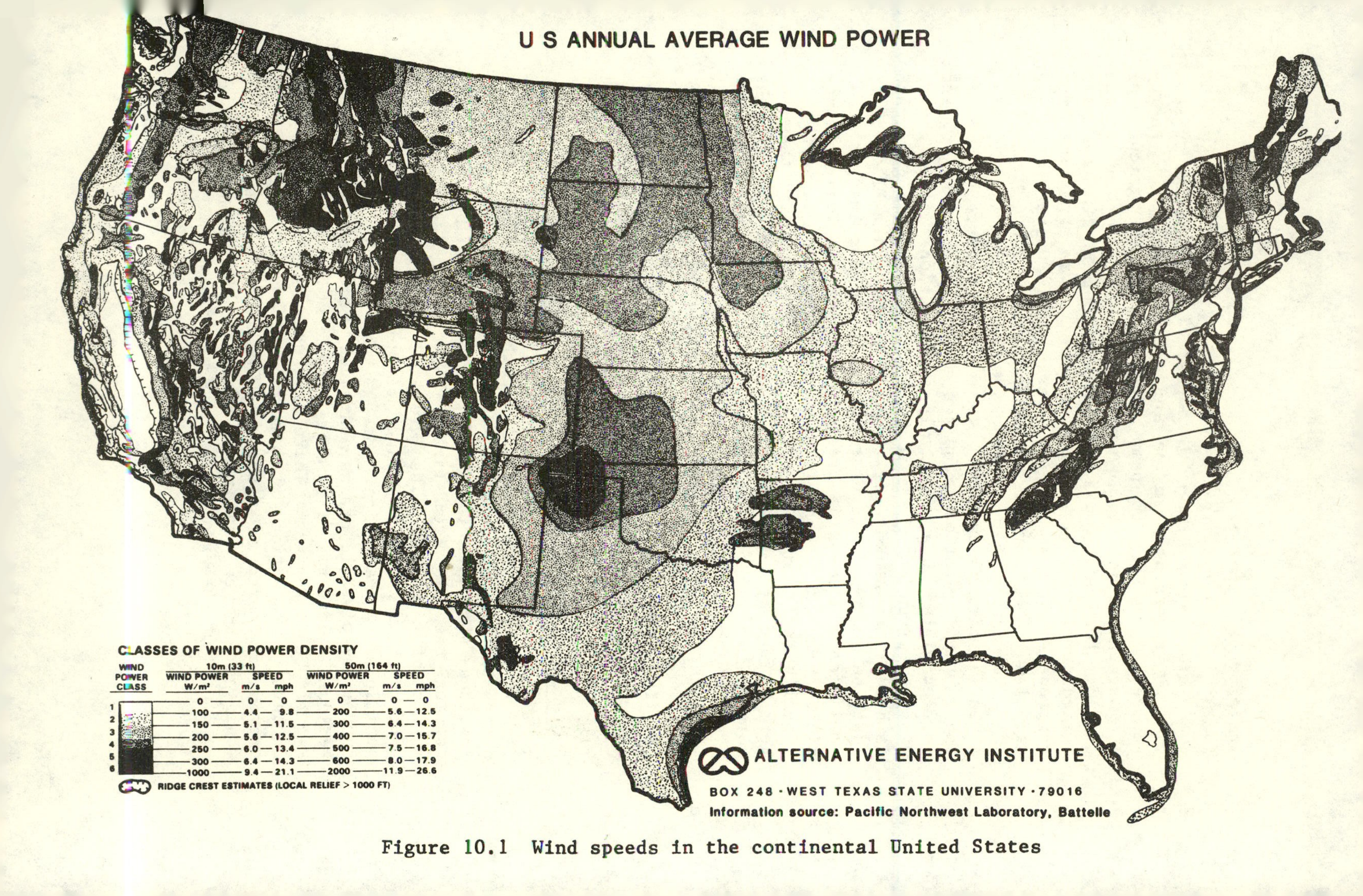

Figure 10.1 Wind speeds in the continental United States

AGRICULTURAL BIOMASS

Agricultural biomass is a possible alternative to some of the increasingly scarce and costly traditional energy sources being used to pump water. Although agricultural biomass can be defined in very broad terms, for the purpose of this discussion it is limited to cotton gin trash (an agricultural waste). Cotton gin trash was emphasized for comprehensive research at Texas A&M University, and detailed economic analysis was conducted.

The cotton gin trash is an especially attractive potential biomass feedstock since it is a waste product of the ginning operation, requires a cost to dispose of, and is accumulated in a relatively few locations (i.e., cotton gins). Currently, the cost to dispose of cotton gin trash is between $5 and $10 per ton on the Texas High Plains. In addition, there is speculation that returning the gin trash to fields exposes wide areas to disease and weed infestations. However, there is also the nagging issue of continually taking organic matter from the land and the long term effect on productivity. Other than recognizing the potentially serious implications of this problem, the productivity issue is not addressed in this report. Thus, although the analysis is based on cotton gin trash, the basic results are applicable to a multitude of biomass materials such as peanut hulls, rice hulls, wheat straw, corn stalks, wood shavings, and other analogous materials.

System Description

To provide some insight into the location of potential agricultural feedstocks for a fluidized-bed gasifer, crop residues by major crops are presented in Table 10.4 and logging residues from timber harvest in Table 10.5 by region of the U.S. The Midwest is important in crop production and thus has large quantities of crop residues available. In addition, some of the West produces large quantities of small grains and rice. Forestry with logging residues indicates a potential feedstock in South and Pacific Coast Regions. There are numerous local areas with intensive production of alternative crops. This would provide a source of feedstock for energy conversion but would require an evaluation based on the unique characteristics of the area and biomass material.

Table 10.4
Estimated crop residues available for energy feed stock

State	Quantity (million tons/yr)
Corn	
Illinois	8.0
Indiana	4.6
Iowa	6.9
Minnesota	4.2
Nebraska	1.8
Ohio	2.6
Small Grains	
California	1.8
Illinois	1.0
Minnesota	6.1
South Dakota	1.8
Washington	3.0
Wisconsin	2.0
Sorghum	
Colorado	0.12
Kansas	0.72
MIssouri	0.28
Rice	
Arkansas	1.9
California	1.1
Texas	1.2
Sugarcane	
Florida	0.53

Source: Office of Technology Assessment.

After a review of relevant literature and site visits throughout the U.S., the biomass for energy team at Texas A&M University identified the fluidized-bed gasification unit as the most likely method of technically and economically converting agricultural biomass to energy. Results reported here emphasize the economic implications of fluidized-bed gasification technology.

Fluidized-bed systems provide several unique operating characteristics which are needed for low-grade biomass

Table 10.5
Estimated logging residues per year

Region	Quantity (million tons/yr)
North	16.0
South	32.8
Rocky Mountain	7.0
Pacific Coast	28.2
TOTAL	84.1

Source: Office of Technology Assessment.

fuels (LePori, et al.). The sand-like particles absorb and store heat while the turbulence of the churning bed maintains a uniform temperature through the bed. When fuel is introduced, rapid and efficient heat transfer occurs from the hot inert particles to the fuel. The large quantity of inert particles relative to fuel particles enable stable temperatures to be maintained. Solid fuels having wide variations in size and moisture can be used in fluidized-beds, so it is an appropriate method to use for biomass fuel conversion. A diagram of a fluidized-bed gasifer-boiler as designed at Texas A&M University is presented in Figure 10.2.

A fluidized-bed system includes an air delivery and distribution system, distributor in the base of a refractory lined chamber. High velocity air enters the chamber through the distributor from below and suspends some inert granular material. The suspended material acts as a fluid with lighter material floating out the top and heavier material sinking. The inert granular material (bed material) distributes and transfers heat to the fuel particles as they are introduced into the reaction chamber. Fuel particles can be coal, or in the example of biomass, cotton gin trash, or sorghum stalks. Since temperatures can be closely controlled and the bed material is efficient in transferring heat to the fuel, the fluidized-bed system is applicable to many feedstocks with widely varying moisture contents.

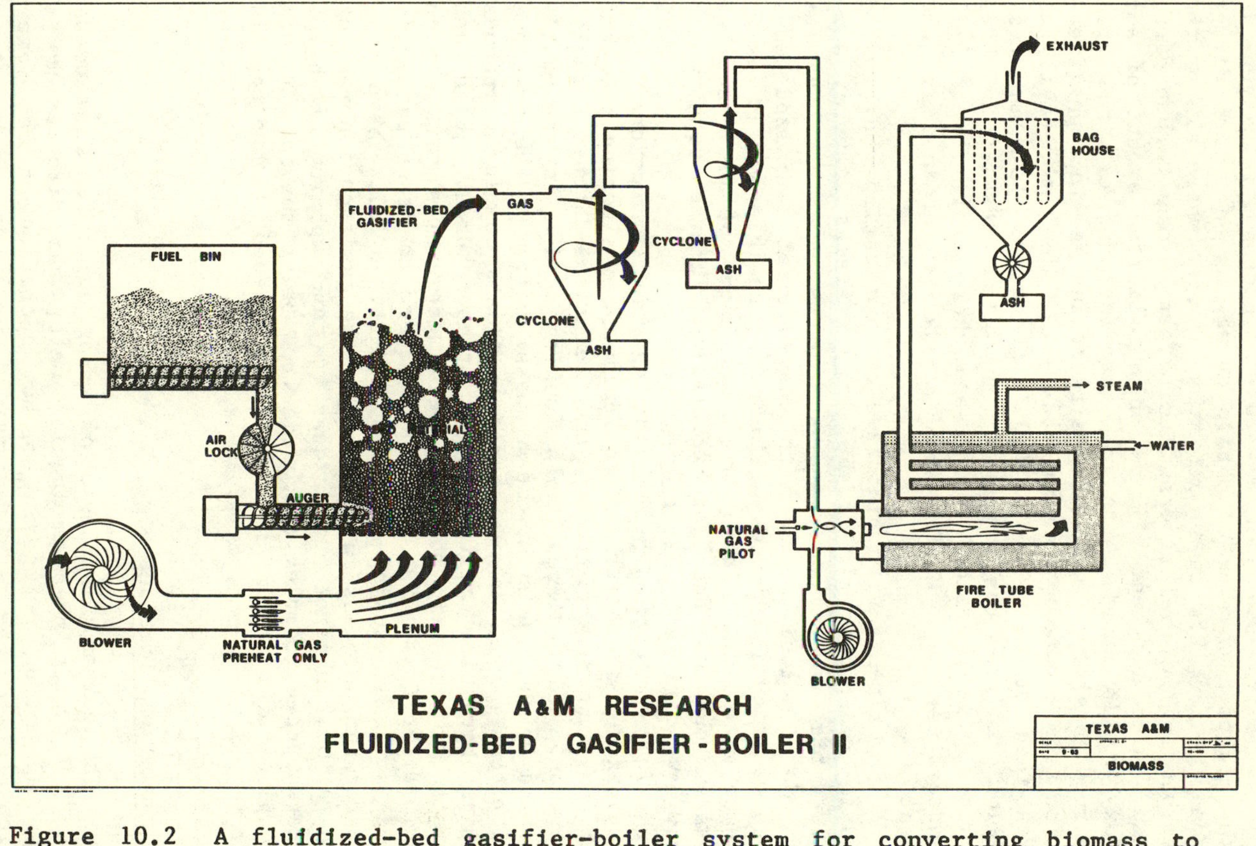

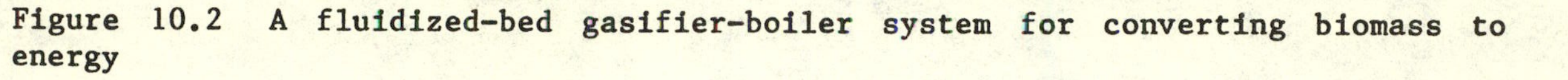
Figure 10.2 A fluidized-bed gasifier-boiler system for converting biomass to energy

Fluidized-beds can be used with excess air (oxygen) in the reaction chamber to obtain energy as heat from direct combustion. This is an oxidation process where all of the carbon in the fuel reacts with oxygen giving carbon dioxide, water and heat. Where only a limited amount of air (oxygen) is supplied to the reaction chamber, the unit serves as a gasifier. Using a fluidized-bed in a gasification mode produces a combustible gas. The combustible gas can then be used as a fuel. If oxygen is completely eliminated from the combustion, pyrolysis results in three types of fuels, liquid, gas, and char. This report will address use of a fluidized-bed system to produce gas and heat.

Experiments with fluidized-bed combustion and gasification of cotton gin trash, sorghum stalks, manure, sawdust, rice hulls, and corn cob biomass materials have led to the conclusion that with conventional fluidized-beds, the gasification process is a more feasible way of using these materials. There are newer types of fluidized-bed systems such as "Fast" or circulating fluidized-beds, which may have different characteristics.

The combustion process using the conventional type fluidized-bed creates severe fouling and slagging problems in the upper regions of the reaction chamber, outlet section, and cyclone. This fouling action is apparently caused by chemicals in the biomass fuel. In the presence of excess oxygen, these chemicals create complex compounds which have low melting point temperatures. High ash content of certain biomass fuels accelerates this action and particles in the gas stream adhere to other particles and surfaces causing buildups. Wall, et al., have discussed this process in detail. In the gasification mode of operation, absence of excess oxygen apparently helps prevent these complex compounds from being formed. A more complete discussion of the fluidized-bed technology is presented in LePori, et al.

Economic Implications

The economic analysis of using agricultural biomass for energy to pump irrigation water was directed to a typical cotton and grain sorghum operation on the Texas High Plains. The results consider two cases, gasification for fueling an internal combustion engine to pump water and installing a boiler for electricity power production to use on electric powered pumps. Using the fluidized-bed gasifier unit to produce combustible gas provides a fuel for a boiler with steam from the boiler driving a turbine generator. For the two types of energy produced, implications for the farm are presented.

Gasification. The economic feasibility assessment of producing a low energy gas from agricultural residues was based on a 640 acre irrigated farm that had four wells and good groundwater (Lacewell, et al., 1981). Although the study area was the Texas High Plains, the coefficients are applicable to many regions of the U.S. The initial objective was to produce sufficient gas using cotton gin trash in a fluidized-bed gasification system to exactly offset irrigation energy requirements of a cotton farm. The analysis was extended to basically year-round operation of the gasification unit. It was assumed that the gin trash is put in modules of 10 tons each. The cost to purchase, module, transport, and store gin trash was taken from Moore, et al.

The feedstock input rate was 1,200 pounds per hour or 14.4 tons per day. The system would be used at about one-tenth of capacity to exactly offset irrigation energy requirements since irrigation does not occur throughout the year (Masud, et al.). This is shown in Column 3 of Table 10.6. For this system, operation costs alone of $2.23 per one million btu (not including any fixed costs) are competitive with current natural gas cost. Because the system would not be used all year long, the total cost is very high, ($7.83 per one million btu). This is compared to a current natural gas price of near $3.50 (Table 10.6) (Masud, et al.).

Column 4 of Table 10.6 indicates the effect of extending use of the fluidized-bed gasifier to 292 days per year. This is 80 percent utilization and approximates the cost of electrical generating facilities (Pollard). During one year the plant requires 4,205 tons of trash or gin trash from about 10,700 acres of cotton production. In this system, gas production beyond irrigation needs would have to be sold for credit or used in other farm activities. The total cost of gas production in this case is much more attractive, at $3.56 per one million btu (Masud, et al.). However, the cost estimate does not include a clean-up charge or power loss for an engine using the low energy gas. Also, even clean low energy gas is unsuitable for pipeline distribution. Costs of the low energy gas are expected to be slightly higher when using grain sorghum or

Table 10.6
Estimated costs of a typical 640-acre irrigated cotton farm using gin trash and on-farm gasification to produce energy[a]

Item	Unit	Offset Irrigation Energy Use	Year-Round Operation[b]
Energy Used for Irrigation[c]	mbtu	5,534.00	5,534.00
Total Energy Produced	mbtu	5,534.00	36,441.60
Fluidized-Bed Gasifier & Automatic Feeder[d]	$	175,000.00	175,000.00
Annual Fixed Cost	$/year	30,975.00	30,975.00
Fixed Cost per Unit Energy[e]	$/mbtu	5.60	0.85
Operation Costs[f]	$/year	12,322.00	98,759.00
Operation Costs per Unit Energy	$/mbtu	2.23	2.71
Total Cost per Unit Energy[g]	$/mbtu	7.83	3.56

Source: Masud, et al.

[a]Assumes a fluidized-bed gasification unit with 8.5 mbtu of gin trash in and 5.2 mbtu of energy in the form of low energy gas out per hour.
[b]Based on 80 percent utilization or 292 days of operation annually (K. Pollard).
[c]Based on the improved furrow irrigation system, otherwise the value is 8,788 mbtu.
[d]Based on bids taken by Texas A&M University and as reported by Levelton and O'Conner.
[e]Based on total production of 5,534 mbtu of gas. For 8,788 mbtu of gas the cost would be $3.53 per one million btu.
[f]Derived from Beck and Parker by scaling down and deleting labor, income tax, and depreciation (part of fixed costs) and setting feedstock cost to $15 per ton) for moduling, handling, transportation and storage (Moore, et al.).
[g]Does not include a distribution cost, gas clean-up cost or cost associated with engine power reduction for the lower energy gas. Thus, costs of gasification are underestimated.

other similar residues due to a lower btu per pound and the need for field baling and hauling (Lacewell, et al., 1982).

Electrical Generation. This section presents an economic evaluation of an electrical generation unit located on-farm. The on-farm electrical generation unit is assumed to include a fluidized-bed combustor, a boiler and a turbine. The size is sufficient to satisfy energy requirements of four irrigation wells of the typical Texas High Plains farm. Surplus electrical generation would be put into a grid system for a credit.

This electrical generation unit requires about 1,100 pounds of gin trash per hour or 3,854 tons a year based on 292 operating days. Total investment for the plant is $275,000. Annual fixed costs would be about $50,000, costs to module, deliver and store cotton gin trash $57,816 and operating costs $39,560. This implies an annual outlay of nearly $150,000 to produce 2,375,700 kwh of electricity. This amounts to a cost of 6.15 cents per kwh (Table 10.7).

Approximately $8,900 of annual electricity purchases are avoided by using the on-farm generation plant, and two million kwh of electricity are available for sale at two cents per kwh. Comparing all revenues (sales of electricity) and credits (on-farm electricity produced and used) to costs of producing the electricity on-farm, the electrical generation plant provides a net loss of $104,171 per year. Of course, adjustment in electricity cost or costs of the system could significantly affect these results (Masud, et al.).

Economic analysis of the most attractive current technology and agricultural biomass feedstocks suggests that traditional fuels are the least expensive source of irrigation power. However, the cost of on-farm production in both the gasification and electrical generation mode are near to the current purchase cost of natural gas or electricity. However, the farmer cannot sell low energy gas or electricity for the rate charged by distributors. Since in both the gas and electricity production cases, surplus energy beyond farm needs is generated, a loss is estimated for on-farm energy production. As irrigation fuel costs continue to adjust, these results are sufficiently encouraging to justify periodic reevaluation.

PLANT OILS

Plant oils represent readily available substitutes for diesel fuel. Since many irrigation wells in the Great

Table 10.7
On-farm electrical generation using cotton gin trash to produce pumping energy, unit size 7.7 mbtu in and 339 kw out per hour[a]

Item	Unit	Value
Electrical Output of Generator	kw	339.00
Equipment Investment[b]	$1,000	275.00
Other Start-Up Cost[c]	$1,000	13.20
Energy Input[d]	mbtu/year	53,961.60
Energy Produced	1,000 kwh	2,375.70
Energy not Used for Irrigation	1,000 kwh	2,094.00
Fixed Cost on Investment	$/year	48,675.00
Feedstock Costs	$/year	57,816.00
Operating Costs[c]	$/year	39,560.00
Total Costs[e]	$/year	146,051.00
Total Costs per Unit Energy[e]	¢/kwh	6.15
Credit for Sale	¢/kwh	2.00
Total Credit[f]	$/year	41,880.00
Net Producer Costs	$/year	104,171.00

Source: Masud, et al.

[a]Based on a 640 acre farm with 4 irrigation wells. It is assumed the system is operated year round using a fluidized-bed combustor, boiler and prime mover to generate electricity. The system size was established as 7.7 mbtu input and produces 339 kw of output per hour.
[b]Based on bids received by Texas A&M University. Includes an allowance for an automatic feeder to the fluidized-bed gasifier.
[c]Derived from Beck and Parker (1979) estimates and scaled down.
[d]Based on 80 percent utilization or 292 days per year (K. Pollard).
[e]Assumes cotton gin trash is purchased, moduled, transported and stored for $15 per ton (Moore, et al.).
Operation and maintenance costs are not well developed.
[f]Includes a credit of $8,982 for irrigation fuel costs provided by the system.

Plains and West are powered by diesel, the economic implications of substituting plant oils is important. Plant oils such as soybean oil, corn oil, cottonseed oil, and sunflower oil can be directly used in diesel-powered irrigation pumps without engine modifications although the long term effects on engines is unknown (Griffin, et al.). The information reported here considers the plant oil crushing capacity of the U.S. and includes an overview of the impact of diverting different amounts of plant oils from their traditional markets (Griffin, et al.).

Oilseed Crushing Capacity

In the short run, crushing capacity may be a limiting factor in substituting plant oils for diesel fuel. Generally, production of oilseed crops trended upward during the 1970s. Annual U.S. production is around two billion bushels for soybeans, ten billion pounds of cottonseed, four billion pounds of sunflower seed, and four billion pounds of peanuts. This translates into 13 billion pounds (1,688 million gallons) of plant oils from these four crops (U.S. Department of Agriculture).

Crushing levels for soybeans, cottonseed, sunflower seed, and peanuts were 65, 8.2, 1.4 and 0.5 billion pounds, respectively in 1980/1981 (U.S. Department of Agriculture). Annual crushing capacity, however, is estimated at 85, 11, 4 and 4 billion pounds for soybeans, cottonseed, sunflower seed, and peanuts, respectively (Griffin, et al.). Excess plant oil crushing capacity in the U.S. suggests sufficient facilities to crush an additional 665 million gallons of plant oils or a total capacity of 2,353 million gallons. The energy content of the surplus capacity of plant oil crushing plants (665 million gallons of plant oils) is equivalent to 624 million gallons of No. 2 diesel fuel. Total estimated diesel use for U.S. field crop production in 1979 was 2.0 billion gallons (Kolmer).

Impact of Diversion

The use of plant oils as substitutes for diesel fuel would influence cropping patterns, crop prices, farmer profit, and consumer well-being. A direct comparison of diesel fuel price and plant oil price indicates that the use of plant oils in diesel engines would cost two to three times as much as diesel fuel on a btu basis. Such a comparison is illustrated in Figure 10.3. Here, the record of market prices for various plant oils is contrasted to historical diesel fuel prices. All of the prices employed in this comparison are wholesale price for bulk quantities at specific locations; taxes have been excluded (Griffin, et al.).

Since plant oils are centrally processed (off-farm), the comparison related in Figure 10.3 is fully germane to the economic potential of plant oils as fuel for irrigation. Only if there are on-farm opportunities to process oilseeds more cheaply will this comparison be inappropriate for the fuel-for-pumping issue. Information concerning the economics of on-farm oilseed processing is sparse. However, a study by Hoffman, et al., (1981) implies that sunflower oil can be produced on-farm at a cost of $2.39 or $4.36 per gallon of diesel equivalent--depending on whether or not labor and building costs are included (Griffin, et al.).

An evaluation of the expected effect of substituting plant oils for 5 and 10 percent of the diesel fuel used in agriculture was done by applying a national agriculture econometric model (TECHSIM). With the diversion of plant oils to fuel use, total planted acres of major crops in the U.S. were estimated to increase between 3 and 6 percent. Acres of oilseed crops increased, especially cotton and soybeans, while acres of corn, small grains, and grain sorghum declined (Griffin, et al.).

Because of the estimated cropping pattern shifts and changes in output, crop prices were affected. The results indicated a price increase for cottonseed and soybeans but a price decline for cottonseed meal, soybean meal, and cotton lint. Corn, small grains, and grain sorghum prices increased in response to the decline in acres and production. With the decline in price of cottonseed meal and soybean meal feed-stuffs, livestock producers increased the numbers of animals fed which depressed fed beef price. Non-fed beef numbers declined; thus, there was some price increase for non-fed beef.

Farm profit was estimated to increase in all production regions. Generally, Corn Belt, Delta States, Central Plains, and Texas producers would be expected to experience the greatest increase in profit if plant oils are diverted to use as a fuel.

However, only field crop producers and final consumers of livestock products are estimated to be beneficiaries of

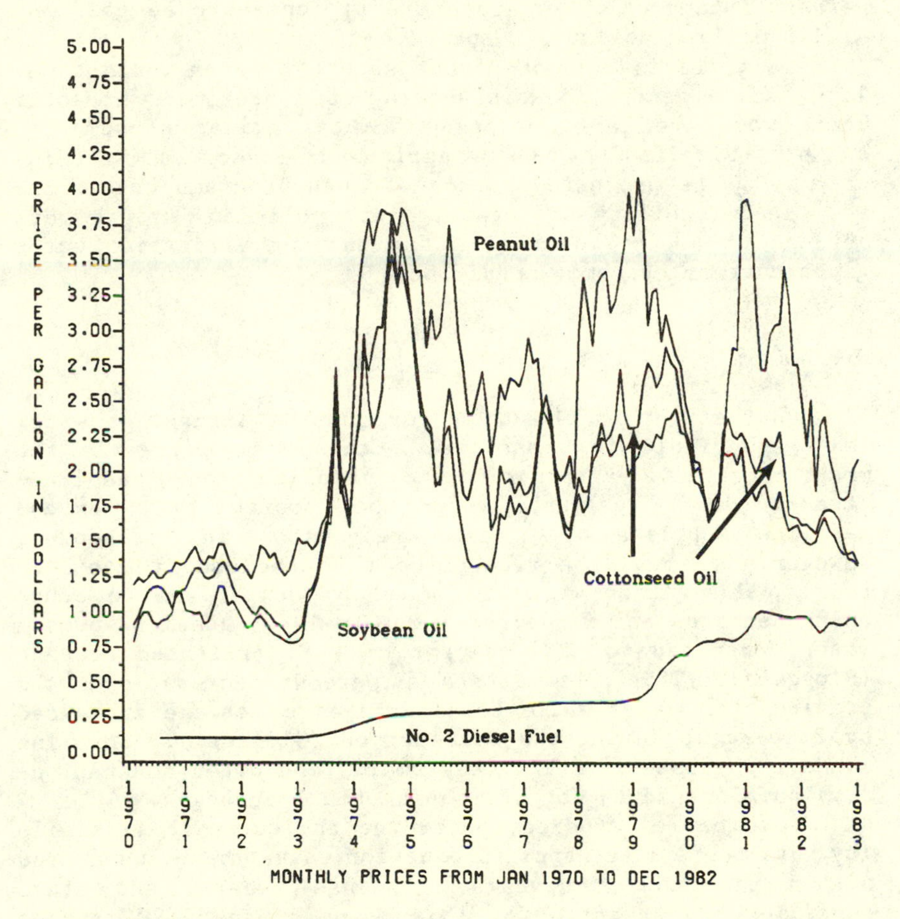

Figure 10.3 A comparison of selected plant oil and diesel fuel prices, 1970-1982

the program. Food processing and marketing sectors, meal and oil using industries, livestock producers, and food consumers all incur a net loss. The total economic impact of substituting plant oils for diesel fuel would be negative as shown in Table 10.8. Total changes in the sums of producers' and consumers' surpluses due to the 5 and 10 percent substitution programs are projected to be -516 and -1,135 million dollars, respectively, in 1990.

Thus, it is estimated that such a program would cause large and generally disadvantageous economic impacts. There would be large cropping shifts, price adjustments, and profit shifts throughout agricultural sectors (Griffin, et al.). The projected losses of such programs underscore the inadvisability of government policies or private actions to employ plant oils for pumping irrigation water given current economic conditions.

ETHANOL

Another alternative fuel for pumping irrigation water is a substitute for gasoline. Ethanol is currently the most attractive alternative to gasoline, and gasoline-ethanol blends (gasohol) offer the opportunity to extend gasoline supplies with minor reductions in horsepower. However, it should be recognized that the use of gasoline to pump irrigation water is small by comparison to other energy sources and is decreasing. In 1980, gasoline pumped water accounted for 2.3 percent of U.S. irrigated acreage (Sloggett). This represents a 33 percent decrease over the previous six years while total irrigated acreage increased by 21 percent during the same period (Sloggett). Gasoline tends to be used in relatively small (and often nonstationary) pumping plants for irrigating small acreages.

A number of studies concerning the economic feasibility of on-farm ethanol production have been conducted within the past ten years. Because federal and state subsidies have existed for 10 percent ethanol (90 percent gasoline mixtures) most research has focussed on this gasohol mix. Recent studies include those of Bowker and Griffin; Dobbs and Hoffman; and Hoffman and Dobbs.

Using experimental data Bowker and Griffin found average total costs per gallon of ethanol to be $4.82 and $1.74, respectively, for 9,000 gal/year and 144,000 gal/year distilling plants. Costs for building, equipment, insurance, taxes, feedstock (sorghum), labor, water

Table 10.8
Welfare impacts of diverting plant oils for use as diesel fuel

Rent or Cost Measure	Five Percent Diversion[a]		Ten Percent Diversion[a]	
	1982	1990	1982	1990
	-------- million dollars --------			
Total Crop Producer Rents[b]	1,914	1,420	3,666	3,112
Total Crop Forward Industry Rents[b]	-987	-788	-1,768	-1,477
Total Meal and Oil Industry Rents[b]	-1,297	-720	-2,565	-1,665
Total Livestock Producer Rents[b]	-63	-56	-104	-80
Total Livestock Wholesale & Retail Producer Rents[b]	-74	-65	-121	-92
Total Livestock Final Consumer Surpluses[c]	140	124	230	173
Total Program Cost of Purchasing Plant Oils	429	431	1,095	1,106
Total Welfare = sum of all Rents Less Program Purchase Cost	-796	-516	-1,757	-1,135

Source: Griffin, et al.

[a]The diversion level represents the amount of diesel fuel btu's in agriculture that are replaced by plant oils.

[b]Rents are the returns to capital investments, operator labor and management.

[c]Improved welfare of the consumer.

ectricity, natural gas, enzymes, yeast, acid, mainte-nce, and repair and by-product credits were considered in this analysis. The use of corn as a feedstock resulted in slightly higher costs. Based on the price of gasoline, the value of state and federal subsidies, and the cost of residual water removal, a net value for the produced ethanol was estimated to be $1.59 per gallon. When these findings were used in a net present value analysis for the larger, more efficient plant size, the net present value was -$289,000 (with a 12 year planning horizon and a 10 percent discount rate).

The results obtained by Hoffman and Dobbs are similar. Average total costs in a 175,000 gal/year facility were determined to be $1.78 per gallon. In their 1983 study Hoffman and Dobbs take a somewhat different approach by assuming that ethanol produced by a single plant will be used to replace 25 percent of the gasoline used in a certain South Dakota county. Focussing on production and transportation costs associated with fuel and by-product deliveries, they again find ethanol production to be an uneconomical enterprise. Depending on the specific assumptions being examined, losses ranged from $.61 to $1.03 per gallon of produced ethanol.

Therefore, ethanol is not an attractive alternative energy source for pumping irrigation water. Even the present structure of government subsidies for gasoline use is unable to alter this situation.

GENERAL IMPLICATIONS

This chapter reviews alternative fuels for powering irrigation wells. The technology or state of the art was described with associated economic implications. Fuels included were solar, wind, biomass, plant oils, and gasohol (ethanol). In all cases, technology is available to use the fuels to power an irrigation well. For solar and biomass, there are several technologies available for converting the fuel to power for pumping.

Solar powered electricity generating plants are currently not cost competitive with traditional fuels. However, with technological advances reducing costs, photovoltaic systems are an attractive alternative in the sunny Southwest for powering irrigation pumps in remote areas. Less than 5 percent of all irrigation pumps are expected to be powered by solar energy by the year 2000.

Wind energy can be used as shaft power or to produce electricity for powering an irrigation well. Electricity generation with wind systems including interconnection with an electricity utility grid shows promise in the higher wind regions, as in the Great Plains. The surplus power not used for pumping water from the wind system was assumed to be sold to the electricity utility. With some current wind systems costing less than $1,000 per kw, using 7.5 cents per kwh electricity and a 3 percent discount rate, the wind systems are economically viable in areas with relatively high sustained wind speeds. Average monthly wind speeds of 12 mph or more satisfy the assumptions for economic viability of wind system. This form of energy will become more attractive as the costs of traditional fuels increase.

Agriculture or other biomass is also a potentially attractive feedstock for generating energy. Fluidized-bed gasification technology was assumed for the analysis. Feedstock can include cotton gin trash, grain stalks, rice hulls, wood residues, and other materials. However, there are unique characteristics of each feedstock that affects the method of operation of the gasifer. Using the system to produce low energy gas indicated costs per one million btu between $3.56 and $7.83. At the lower end with full utilization of the system, the cost is competitive with natural gas. However, the low energy gas cannot be stored, sold or fed into gas lines hence must be used as it is generated on or near the production site. In most cases this is not feasible, hence, the system is run at less than capacity substantially increasing the cost of gas generation.

Using the low energy gas to produce electricity results in an estimated cost of 6.15 cents per kwh which is competitive with current commercial rates of electricity. However, much of the electricity generated cannot be used on-farm and must be sold to the utility for much less the the 6.15 cents per kwh. Thus, the system incurs an economic loss over the year. This means that for the use of biomass to be economically competitive, the system must operate at near capacity and all of the energy produced used on-farm.

Plant oils and ethanol (gasohol) are both substitutes for internal combustion engines (i.e., diesel and gasoline, respectively). Ethanol is mixed with gasoline and plant oils are blended with diesel fuel. Long-run implications on engine operation and maintenance are not fully established. However, in both cases, the costs of producing the fuels per unit of energy are significantly greater than for traditional fuels. Thus, only in the case of shortage or large subsidies would these fuels be economically competitive for pumping irrigation water.

In general alternative fuels are not economically competitive with traditional fuels for pumping irrigation water. As technology progresses and relative costs shift, this relationship may change. Thus, it is imperative that study of the role and viability of alternative fuels continue.

REFERENCES

Alternative Energy Institute. "A Brief Discussion of Wind Energy in Texas." West Texas State U., pamphlet, October 1980.

Beck, S.R. and H.W. Parker. "Assessment on Energy from Biological Processes." Task V, Part 1: Engineering Aspects of Thermochemical Conversion, Draft Report to Office of Technology Assessment, U.S. Congress, Dept. of Chemical Engineering, Texas Tech U., 1979.

Bowker, James M. and Ronald C. Griffin. "The Economic Feasibility of Small-Scale Ethanol Production in Texas." Center for Energy and Mineral Resources, CEMR-M56, Texas A&M U., July 1983.

Buzenberg, R.J., M.L. David, J.P. Wagner, E.F. Glynn, G.L. Johnson, and J.K. Shultis, "Market Potential for Wind Energy Applications in Agriculture." Wind Energy Applications in Agriculture, ed. H.H. Kluter and L.H. Soderholm, U.S. Dept. of Agr., NTI-Conf-7905/109, pp. 22-41, May 1979.

Clark, R.N. and A.D. Schneider. "Irrigation Pumping with Wind Energy." Transactions of the ASAE, 23(1980): 850-853.

Dobbs, Thomas L. and Randy Hoffman. "Small-Scale Fuel alcohol Production from Corn: Economic Feasibility Prospects." South Dakota Agr. Exp. Station B-687, June 1983.

Gipe, Paul. "Lone Star Wind: An Overview of Current Wind Energy Activities in Texas." Wind Power Digest, Fall 1979, pp. 36-48.

Griffin, R.C., R.D. Lacewell, G.S. Collins, and H.C. Chang. "The Economics of Plant Oil Substitution for Diesel Fuel." Center for Energy and Mineral Resources MS-7, Texas A&M U., 1983. (Preliminary results were presented in an earlier paper by Collins, G.S., R.C. Griffin and R.D. Lacewell, "National Economic Implications of Substituting PLant Oils for Diesel Fuel." Proceedings of the International Conference on Plant and Vegetable Oils as Fuels. ASAE, Fargo, North Dakota).

Gunkel, W.W., R.B. Furry, D.R. Lacey, S. Neyeloff, and T.G. Porter. "Development of Wind-Powered Heating System for Dairy Application." Wind Energy Applications in Agriculture, ed. H.H. Kluter and L.H. Soderholm, U.S. Dept. of Agr., NTI-Conf-7905/109, pp. 124-157, May 1979.

Hardin, Daniel C. and Ronald D. Lacewell. "Breakeven Investment in a Wind Energy Conversion System for an Irrigated Farm on the Texas High Plains." Texas Water Resources Institute TR-116, October 1981.

Heid, W.G. Jr. and W.K. Trotter. "Progress of Solar Technology and Potential Farm Uses." USD/ERS, Agr. Econ. Report 489, Washington, D.C., pp. 42-51, 76-90, September 1982.

Hoffman, Randy and Thomas L. Dobbs. "A Small Scale Plant: Costs of Making Alcohol." South Dakota Agr. Experiment Station B-686, September 1982.

Hoffman, V., D. Kaufman, D. Helgeson, and W.E. Dinusson. "Sunflower for Power." North Dakota Coop. Ext. Service, North Dakota State U., April 1981.

Kolmer, W. "The Energy Requirements of U.S. Agriculture." Monthly Energy Review, U.S Dept. of Energy, 1979.

Lacewell, R.D., E.A. Hiler, S.M. Masud, and R.D. Kay. "Assessment of Integrated Agricultural and Energy Production Systems." Center for Energy and Mineral Resources MS-4, Texas A&M U., 1982.

Lacewell, R.D., C.R. Taylor, and E.A. Hiler. "Energy Generation from Cotton Gin Trash: An Economic Analysis." Center for Energy and Mineral Resources, Monograph Series, Texas A&M U., 1981.

Lansford, Robert R., Raymond J. Supalla, James R. Gilley, and R. Nolan Clark. "Economics of Wind Energy for Irrigation Pumping." Technical Completion Report submitted to USDA-SEA, by Southwest Research and Development Co., Las Cruces, New Mexico, July 1980.

Larson, D.L. "Utilization of an On-Farm Solar Powered Pumping Plant." Transactions of the ASAE, 22(5): 1106-1109, 1114, 1979.

Larson, D.L. "Operation Evaluation of a Solar Power Plant for Irrigation Pumping." Paper No. 82-2523 presented at the ASAE Winter Mtg., Chicago, December 1982.

Larson, D.L. "Final Report of the Coolidge Solar Irrigation Project." Submitted for publication by Sandia National Laboratories as a DOE research report, July 1983.

LePori, W.A., C.B. Parnell, Jr., R.D. Lacewell, T.C. Pollock, and R.B. Griffin. "On-Site Energy Production from Agricultural Residues." Contract completion report prepared for the Texas Energy and Natural Resources Advisory council, available from Dept. of Agr. Engineering, Texas A&M U., August 1983.

Levelton, B.H. and D.V. O'Conner. "An Evaluation of Wood-Waste Energy Conversion Systems." B.H. Levelton and Associates, Ltd., Printed by Environment Canada, Western Forest Products Lab., 6620 Marine Drive, Vancouver, B.C. VGT 12.

McNelis, B. (ed). Proceedings of Conf. on Solar Energy for Developing Countries, Refrigeration and Water Pumping, UK-ISES, 19 Albemarle St., London W1X 3HA, UK, January 1982.

Masud, S.M., R.D. Lacewell, and E.A. Hiler. "Economic Implications of Cotton Gin Trash and Sorghum Residues as Alternative Energy Sources." Energy in Agriculture, Elsevier Scientific Publishing Company, Amsterdam, pp. 267-280, 1983.

Moore, D.S., R.D. Lacewell, and C. Parnell. "Economic Implications of Pelleting Cotton Gin Trash as an Alternative Energy Source." Texas Agr. Exp. Station B-1382, Texas A&M U., 1982.

O'Brien, W.F., H.L. Moses, D.H. Vaughan, J.A. Schetz, and T.A. Weisshaar. "Application of Windmills to Apple cooling and Storage." Wind Energy Applications in Agriculture, ed. H.H Kluter and L.H. Soderholm, U.S. Dept. of Agr., NTI-Conf-7905/109, pp. 158-171, May 1979.

Office of Technology Assessment. "Energy from Biological Processes, Vol. II - Technical and Environmental Analysis." U.S. Congress, U.S. Printing Office, Washington, D.C., September 1980.

Pham, H.Q. and L.D. Jaffe. "Heat Engine Development for Solar Thermal Power Systems." Proceedings American

Society of Mechanical Engineers Solar Energy Div. Mtg., Reno, NV, pp. 654-659, April 1981.

Pollard, K. Dept. of Agr. Engineering, Texas A&M U., Personal Communication, 1982.

Sir William Halcrow and Partners. "Small-scale Solar-Powered Irrigation Pumping Systems Technical and Economic Review." UNDP Project GLO/78/004, The World Bank, September 1981.

Sloggett, Gordon. "Energy and U.S. Agriculture: Irrigation Pumping, 1974-80." USDA/ERS, Agr. Econ. Report Number 495, December 1982.

Stoiaken, Larry. "The Small Wind Energy Conversion System Market." In Alternative Source of Energy, pp. 20-21, September/October 1983.

Sullivan, N.W., T.L. Thompson, P.E. Fischbach, and R.F. Hopkinson. "Management of Solar Cell Power for Irrigation." Transactions of the ASAE, 23(4): 919, 1980.

Sullivan, N.W. and P.E. Fischbach. "Operational Problems Confronting Photovoltaic Use in the Agricultural Environment." Paper No. 82-2524 presented at the ASAE Winter Mtg., Chicago, December 1982.

U.S. Department of Agriculture, Economic Research Service. "Fats and Oil Situation." Selected Issues 1979-1981, Washington, D.C.

U.S. Department of Commerce. Local Climatological Data, National Oceanic and Atmospheric Administration, Environmental Data Service, Washington, D.C., monthly.

Wall, C.J., J.T. Graves, and E.J. Roberts. "How to Burn Salty Sludges." Chemical Engineering, 14: 77-82, April 1975.

Williams, D.W., D.L. Larson, and R.J. McAniff. "Load Management of On-Farm Solar Electric Power Plants." Paper No. 7 79-3003 presented at the ASAE, Summer Mtg., Winnipeg, Canada, June 1979.

11
Operating with Energy Shortages

William J. Chancellor
Warren E. Johnston

INTRODUCTION

The fate of many existing irrigation systems and the form of all future irrigation systems depend on the cost and availability of fossil fuels and electrical energy. In 1972, 44 percent of the energy required to produce and deliver the products of California's irrigated agriculture to the farm gate, was consumed in water pumping, an example of the energy intensity of western irrigated agriculture. While this pumping energy constitutes only slightly more than 1 percent of the total commercial energy consumed in the state, the availability of irrigation water is of crucial importance to crop production in California, especially for high-valued fruit, vegetable, and specialty crops. In this chapter we first examine several studies that attempt to evaluate the impact of reduced supplies or increased costs of selected energy inputs in agricultural production and, second, make an analogy between a recent water drought and a possible future energy shortage for implications about how farmers respond to a shortage of a crucial input. In the first section, quantitative models of agricultural production systems are used to examine the responses to impose energy price changes or energy availability constraints. The second section makes observations about what happened during the California drought of 1976-77, providing an opportunity to examine the impact of a major curtailment (53 percent in 1976 and 78 percent in 1977) in the supply of surface runoff water in a state where surface water normally constitutes about 60 percent of the irrigation water supply. Both approaches will be used to reflect on impacts of energy price increases or energy use curtailment on how agricultural institutions and managers will allocate energy among its uses and the resulting changes in agricultural production.

AGRICULTURAL ENERGY ALLOCATION MODELS

Four studies are reported here which consider the case of energy curtailments to the agricultural sector in general and then examine farmer responses in terms of total production, the selection of cropping enterprises, and resulting demands for various inputs. In particular, the impact of energy shortages on irrigation water inputs and accompanying changes in economic measures of farm productivity were examined.

Linear Programming Model of U.S. Production

Dvoskin and Heady (1976) present results from a large model with over 100 regions describing production agriculture in the United States. They used linear programming to obtain (for conditions projected in 1985) minimum cost for production and transportation of a fixed level of agricultural crop production under various scenarios: (1) a base model with 1972 cost of production figures and 1974 energy prices, (2) a minimization of 1985 energy use for agriculture, (3) a state in which 1985 energy use for agriculture was 90 percent of the base model, and (4) a minimization of agricultural production costs in the presence of high energy prices.

With 10 percent reduction in energy use (Case 3), there was a 41 percent drop in irrigated acreage from that of the base model; water use nationwide decreased 36 percent. The majority of the irrigated acreage reduction was in the western states. In the Pacific Northwest, irrigation was discontinued on nearly 90 percent of formerly pump irrigated acreage, while the Southwest had only a 10 percent reduction. While energy is not likely to be efficiently reallocated nationally even in a crisis, the study suggests that irrigation in the Pacific Northwest is nearer the margin in terms of national economic efficiency than that in the Southwest.

The study also permits observation about the "shadow price" (profit change per unit change of an input) of certain energy inputs in U.S. agricultural production. The average shadow price for all fuel and electricity inputs was 384 percent of base level prices of these inputs used in the model. The shadow price of irrigation water in the 10 percent energy reduction model, however, rose only 14 percent from the base model to $10.59 per acre-foot on a national basis. The implications of this relatively low increase may be that water in its current use in U.S. agriculture is not tightly constrained relative to the magnitude of change examined (10 percent reduction), so that foregone profits were relatively small. The implications of the high shadow prices for energy sources will be discussed later, as the other models investigated had similar findings.

Dvoskin and Heady also discussed possible impacts and interactions between model results and the domestic and international market demand for agricultural products, with commodity prices strongly affected by reduced supplies. They found that accompanying the 10 percent reduction in energy use, there was a 55 percent increase in commodity prices resulting in a 52 percent increase in gross U.S. farm income. However, because of the impact of reduced energy availability on irrigated crop production, gross farm income increased only 2 percent in the Southwest. And in the Northwest, gross farm income was estimated to decrease by 8 percent.

Quadratic Programming Model of California Crop Production

Adams and Adams, King, and Johnston constructed a quadratic programming model of California agriculture which included demand elasticities for field-crop or vegetable-crop commodities. The model attempted to maintain a realistic balance among commodities such as might be established by farmers responding to economic forces. Physical energy constraints were applied at 20 percent and 40 percent reduction levels yielding production extremes for annual crops in each of 14 regions in California under various cost and risk scenarios. Adams, King, and Johnston examined the impact of constraining fuel allocations regionally, according to some of the historically based emergency fuel allocation programs proposed at that time. Results were then compared to the case in which the same total amount of fuel was allocated to California without localized (regional) constraints.

Shadow prices of constraining inputs (including irrigation water in some cases) were determined. It was found that 20 percent and 40 percent reductions in fuel and nitrogen fertilizer energy inputs to California agriculture resulted in 8 percent and 24 percent decreases, respectively, in the value of annual crops produced. The average, over both reduction levels, of the diesel fuel shadow price was $11.66 per gallon indicating that fuel constraints resulted in very serious reductions of agricultural product value. With the 40 percent reduction in fuel and fertilizer inputs, cropping patterns were shifted to conditions requiring less irrigation water. In a number of individual regions, production was constrained by water availability for certain crops, even though reductions in water availability were not specified in the model. With the 20 percent reduction in fuel and fertilizer inputs, the shadow price for water in regions where it was a constraining factor, was $112 per acre-foot (compared with the base level price of about $6). With a 40 percent reduction in fuel and fertilizer inputs, the shift in cropping pattern to lower-water-use crops lowered the shadow price for water in water-constrained regions to $56 per acre-foot.

Applying the constraints to each region equally and disallowing interregional transfers, Adams, King, and Johnston found decreases in consumer benefits of $139.1 million and producer profits of $124.5 million. These losses resulted from the regional constraints allowing production of only 67 percent of the annual crop acreage and 77 percent of the annual crop tonnage that could have been produced with the same statewide energy cut but with interregional energy transfers permitted. It is expected that if reduced allocations for water pumping energy or for irrigation water were imposed on a regional basis, similar impacts on production would occur--much more costly consequences than when energy and/or water transfers are freely permitted among agricultural producing regions and sectors.

Systems Analysis of California Crop Production

Williams modeled California crop production with the amounts of five selected inputs--tractor power, combine capacity, tractor fuel, fertilizer, and irrigation water--

as independent variables in 45 equations, to predict the regional production of nine major crop categories in five regions of the state. For each of the 45 crop-region cases, the level of input use was optimized holding one at a time of the five inputs at 75 percent or 50 percent of its normal value. The optimization was based on maximization of cash sales minus cash costs.

In California about 44 percent of the energy used for farm inputs is associated with the supply of irrigation water. Williams, in addition to Williams and Chancellor, found that reduction of irrigation water inputs to 75 and 50 percent of unrestricted levels resulted in larger impacts on agricultural production than did reductions in any of the other four energy-related inputs. At 75 percent of normal irrigation water inputs, the group of equations predicted a 15 percent drop in value of statewide agricultural production; at 50 percent, a 42 percent drop in production was predicted. The associated declines in cash profits per acre-foot of water removed from the system were $69 and $104, respectively, in 1969 prices, when actual prices of irrigation water were typically in the $5 to $10 per acre-foot range. Alfalfa was reduced 60 percent in production by a 50 percent cut in irrigation water--while tree fruit and nut production were reduced only 39 percent.

When the input of fuel for field operations was reduced to 75 percent of its unrestricted level, the optimum cropping pattern still called for full applications of irrigation water. Only when tractor fuel was cut to 50 percent of its high production level was use of in irrigation water reduced, and then only for sugarbeets and vegetable crops. The shadow prices of diesel fuel under the 75 and 50 percent availability scenarios were $3.31 and $5.95 per gallon, respectively, compared to the energy price used in the model of $0.185 per gallon of diesel fuel equivalent.

Linear Programming Model of Oregon Agricultural Production

Chen and Wensink modeled the Oregon agricultural production system using linear programming to alternatively maximize cash sales minus cash costs (net farm income), digestible energy production, and digestible protein production. Constraints on diesel fuel and gasoline--independently and in combination--were ranged from 10 to

100 percent of normal supply levels. The model was applied to the agriculture of Oregon's Willamette Valley.

In this model, irrigated acreage was used mainly for vegetable production. When net farm income was maximized with constrained fuel inputs, irrigated acreage did not decrease until diesel fuel availability reached 35 percent of its full-production level. Because of the very high profitability of irrigated vegetable crops, irrigation was maintained even with greatly decreased fuel inputs. Major acreage reductions did take place more quickly for the less profitable grain crops as fuel availability was reduced. Thus, energy use reductions came from three sources: (1) field operations in nonirrigated crops, (2) field operations in irrigated crops, and (3) water pumping for irrigated crops. All three sources would be under competitive pressures to reallocate fuel uses in the face of fuel reductions. Those with the strongest economic return per unit fuel used would be least likely to reduce fuel use. The final result would depend on substitutability between field operations fuels (mainly diesel fuel) and irrigation pumping fuels (electricity, natural gas, and diesel fuel). In the Pacific Northwest, hydroelectric power is the base source for many electric utilities with varying amounts of both oil and natural gas supplying the remaining energy for electricity generation. Also many industries and large institutions can fire their boilers alternately with diesel-type oil or natural gas. Thus some petroleum fuels and electricity remain substitutable. This substitutability does not occur at the farm level, necessarily, but is important at the state level (Dvoskin and Heady).

Summary Models Review

In all the models there were sizable ratios between the shadow prices for energy or water supplies and the then current costs of those supplies. These ratios for energy ranged from 3.84 (Dvoskin and Heady) to 42.5 (Chen and Wensink). Ratios on the order of 15 were found for irrigation water which is the main agricultural consumer of energy in most irrigated agricultural areas. There are some possible implications of these results. Either farmers are not using energy or water to the maximum profit point where marginal cost equals marginal return to the input, or there is an abrupt discontinuity in the relationship between water or energy inputs and agricultural returns. It is also possible that the models are not sufficiently detailed to reflect the sorts of production technology adjustments and substitutions that farmers and agriculturally related firms and institutions can make to mitigate potential decreases in farm profits when water or energy, or both, are constrained.

It is likely, however, that much more modest increases in water and fuel prices, than those indicated by the shadow price-cost ratios, would be effective in bringing about significant reductions in the use of constrained energy resources for lifting and applying irrigation water. Wherever opportunities exist for the substitution of surface water for groundwater, the use of low energy irrigation technologies, the substitution among fossil fuel types, the mobilization and substitution of renewable energy resources for fossil fuels, or water and fuel conservation in which disadvantageous economic consequences are small, behavior will be much more responsive to fuel curtailments than is indicated by these models.

ACTUAL RESPONSES OF FARMERS TO SUPPLY SHORTAGES

In irrigated agriculture, energy use for lifting, moving and applying water is frequently the largest single item in energy budgets for on-farm production. Physical shortages of energy, therefore, could result in reduced amounts of irrigation water applied. The two consecutive drought years in California, 1976 and 1977, afforded an opportunity to obtain information about possible farmer responses to curtailment in the amount of irrigation water. That is, the drought experience may have some relevance to a future energy-caused water shortage. In the 1976 water-year, precipitation was 65 percent of normal; surface runoff was 47 percent of normal, and in 1977, precipitation was 45 percent and surface runoff 22 percent of normal (Department of Water Resources).

In summarizing numerous studies, Davenport, et al. concluded that crop yield usually declines in direct proportion to the decrease in the amount of water evaporated and/or transpired by the crop. In a state where surface runoff constitutes about 60 percent of the total water supply, and irrigation accounts for approximately 85 percent of the water consumed, farmers and water organizations have real concerns about major reductions in water supplies and associated potential crop losses. What farmers and water organizations did during the drought years, therefore, seems worthy of examination in light of the very serious economic consequences forecast by state officials, farmers, and business persons. When the drought ended, however, impacts on the value of California crop production were not nearly as serious as predicted; in fact, in overall value, the effect was almost insignificant.

The comparison of the drought experience with that of a fuel-supply-induced pumped water shortage has certain limitations. First, no one expected the drought to last longer than three years whereas energy shortages could be prolonged or even permanent. Thus, adjustments during the drought are more aptly characterized as short-run rather than long-run in nature. Second, major investments in surface water facilities had been made over the years specifically to cope with temporary droughts. Buffer supplies of energy, however, are not available or would require a long period of development. Third, groundwater pumping, as a substitute, relieved the surface water shortage whereas in the case of an energy shortage, there may be no such alternative. An energy shortage could mean that water movement operations in the surface water system would also be constrained.

The most severe effects of the drought fell on those most dependent on rainfall sources, in particular, livestock operations dependent on range or nonirrigated pasture feed sources. For these operations there were few alternatives other than reduction in livestock numbers or purchasing feed. Dryland barley and wheat farmers either didn't plant or planted only the least drought-affected lands. However, these rain-dependent type farming operations are not relevant to the discussion since their water supply would not be affected by fuel or electricity shortages.

Tables 11.1 and 11.2 illustrate two of the three main methods used in coping with the drought: increased groundwater pumping and extreme reservoir draw-down. The third method, transferring water from one zone to another, also usually involved pumping.

The heavy reliance on groundwater pumping with significant overdrafting of groundwater basins during the drought can be seen in Table 11.1. The extent of overdraft could be tolerated only as a short-term solution; many years of careful management will be required to bring groundwater

Table 11.1
Central Valley water use--millions of acre-feet[a]

Source	1975[b]	1976	1977	1978[b]
Surface	15.25	12.79	8.97	15.77
Ground	11.65	15.84	18.29	10.74
(Overdraft)	(1.34)	(3.67)	(5.57)	(3.60)
TOTAL	26.90	28.63	27.26	26.51

[a]The Central Valley contains 75 percent of California's irrigated acreage.
[b]1975 and 1978 were nondrought years.

Table 11.2
Water storage in ten major California reservoirs

		Storage		
Item	Capacity	Nov. 1 1976	Nov. 1 1977	Nov. 1 1978
Millions of Acre-Feet	17.52	6.93	2.71	16.30
Percent Capacity	100.00	40.00	15.00	93.00

Source: Piper.

levels in these basins back to their 1975 levels. In the case of an energy shortage such overdrafts would not likely occur.

During the drought about $300 million were spent by farmers in drilling new wells and improving old ones. Figure 11.1 illustrates the energy related trends for Northern California where 80 to 90 percent of agricultural electricity sales are used for water pumping. Despite the

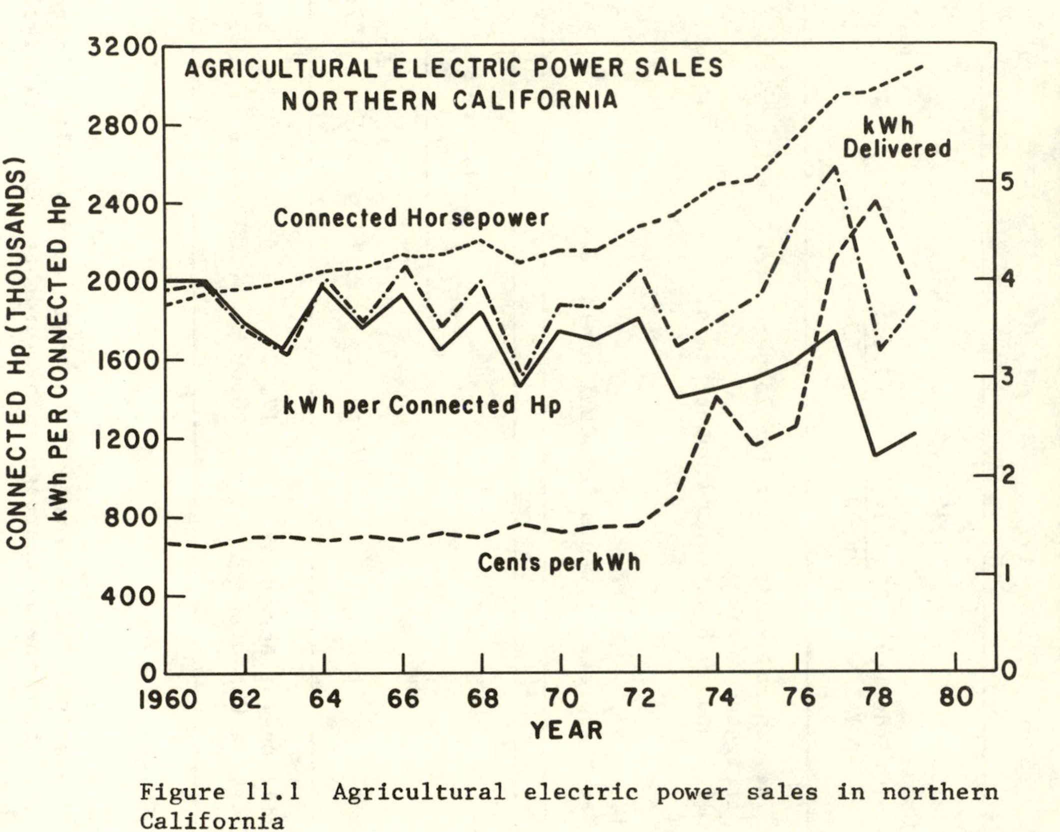

Figure 11.1 Agricultural electric power sales in northern California

fact that energy prices rose sharply during the drought years, more pumping was connected and the annual hours of operation increased resulting in a sharp increase in total electrical energy use. Abundant rainfall in 1978 resulted in a 40 percent drop in agricultural electrical energy use from its 1977 level.

Table 11.1 also indicates that additional surface water development could be a potential alternative to extensive groundwater pumping. Surface water projects that do not require pumping constitute a low-energy alternative in the supply of irrigation water (Roberts and Hagan). However, nearly all low-cost development alternatives have already been thoroughly exploited leaving little potential for such future development. Many regard future reservoir development to be economically infeasible.

Table 11.2 shows the great usefulness of California's water reservoirs in meeting drought-year water needs. Although surface runoff water was 47 and 22 percent of normal in 1976 and 1977, respectively, reservoir storage allowed surface water use to drop by only 17 and 42 percent, respectively. The result suggests that analogous storage of fossil fuel energy could serve to meet short-term energy shortages. But, the high cost of the fuel inventory (including the cost of the fuel, interest, and holding costs) would have to be recaptured--perhaps from those benefiting most directly from such insurance.

In 1977, the California Drought Task Force formulated drought damage forecasts for three different weather-energy scenarios, an optimistic case, the most likely case, and a pessimistic case (Department of Water Resources). In the optimistic case energy for pumping was assumed to be freely available, in amounts needed. In the pessimistic case, less energy was assumed available than in 1975. Forecasts for drought-caused reductions in gross agricultural sales for 1977 were $450 million (5 percent) and $1,364 million (16 percent), respectively, for the optimistic and pessimistic cases. In the optimistic case where energy was sufficient, 94 percent of the loss was related to reduced livestock sales. But in the energy constrained case, only 36 percent was livestock related. The rest was in crop loss, indicating the importance of energy availability to the impact of water shortage on California's agricultural economy.

Table 11.1 also shows that despite the drought, the total use of water in the Central Valley was actually higher in 1976 and 1977 than in the normal water-years of

1975 and 1978. Although a considerable amount of water was pumped, transferred, or taken from reservoirs, much was not available at the most crucial locations and times. Thus, only when water supplies and drought management techniques were examined at local and county levels, was there any evidence of actual shortages and the methods used to cope with them.

Techniques used to cope with localized irrigation water shortages involved reduced demands on primary water supply sources. They are applicable to energy-shortage situations in which reduction in energy for water pumping constitutes the main cause of water shortages. The main strategies to cope with the drought were applied on a field-by-field basis. The three most common approaches were:

1. Not supplying irrigation water to acres normally irrigated. Some of this reduction in irrigated acreage was planted in dryland crops; the balance was not planted.
2. Reduction in the acreage of high-water-requirement crops, such as rice and sugarbeets. Instead, more low-water-requirement crops such as cotton, were grown.
3. Using less than normal water on irrigated crops and sustaining some losses of yield or quality. Yield losses were usually less severe than expected because of unusually favorable weather conditions and the use of effective water-conserving application techniques. In some areas, the rain that did fall came just in time for crop establishment, minimizing the need for supplemental irrigation water applications.

Other methods used to cope with the drought included county wide water conservation management programs, reducing acreage in irrigated pasture, switching from surface application methods to sprinkler or drip methods, establishing tailwater return systems and using drainage water for irrigation, winter preirrigation (considerable cotton acreage was harvested during the drought with little more than a preirrigation), rationing water using various allocation criteria, and abandonment of double cropping. For the Central Valley of California these adjustments resulted in major irrigated crop acreage reductions in only a few crops in 1977: grain sorghum production, 34 percent of the 1973-75 average; safflower, 59 percent; rice, 68 percent; sugarbeets, 79 percent; and lima beans, 80 percent.

Yields in the Central Valley during 1977 were noticeably below the 1973-75 average for only a few crops: olives, 55 percent of the 1973-75 average; wine grapes, 89 percent; and raisin grapes, 90 percent. Crop prices in 1977 were generally high, due in part to drought-caused production shortages. The resulting total value of irrigated Central Valley crops in 1977 was 116 percent of the 1973-75 average; and 1977 was a record year in terms of gross value for California agriculture and a second-to-record year for net farm income.

SUMMARY OF DROUGHT-ENERGY SHORTAGE ANALOGY

It appears that the impacts of sudden water supply changes can be mitigated as man-developed facilities, institutions, and innovation play larger roles in a given system. Recall that systems in which man's role is less intensive--livestock grazing and dryland cropping operations--suffered the most during the drought. The juxtaposition of record drought and record farm product value focuses attention to farmers' and farm-related agencies' capacity to successfully innovate in the face of adversity. Two main and complementary approaches were used. Actions were designed to first get the water from somewhere and then to produce the most valuable crops possible from available water.

Operations used in obtaining water were so effective that more total water was supplied during the drought years than was used in normal years. This alternative would not likely be applicable to situations in which reduced water availabilities were caused by energy shortages. But farmers could be expected to respond to an "energy drought" using similar strategies so several aspects of solutions to the water drought might yield useful parallels in dealing with energy drought.

Background data on rainfall probabilities, groundwater resources, and water requirements for crops were extremely important in forming strategies for successfully dealing with the drought and in recognizing that it was a short-term problem. The available reservoirs had drought-mitigation benefits among the justifications for their construction. These man-made supply buffers together with the natural groundwater supply buffers, had a capacity about equal to one year's total water needs. The willingness of representatives from many sectors to come together and work toward a common goal, made water transfers feasible and encouraged persons throughout the farming and water industries to work toward solutions rather than competing for available water. Furthermore, the knowledge that the near future would, in high probability, bring improved rainfall allowed water suppliers and users to enter into ad hoc short-term arrangements without the detailed considerations necessary for longer term agreements.

There are important dissimilarities between droughts and energy shortages. For one thing energy shortages are more likely to be of longer duration or even permanent, with little prospect of future alleviation. Another is that energy use for water pumping makes up only a small percentage of total commercial energy use, whereas agriculture accounts for more than 85 percent of water consumed. Conservation on the part of other energy users, therefore, represents for water pumpers, a major resource for sustaining the amounts of energy available for irrigation water purposes. In this competition, political, economic and physical factors are all likely to have bearing on allocative outcomes.

Many of the approaches used during the drought to cope with water shortages would be applicable when water availability was curtailed by energy shortages. The effectiveness of agriculture's response during the drought shows that it is possible to change the technological coefficients of agricultural production. The diverse nature of these water savings approaches and the local variation in their application indicates that in the case of an energy shortage for water pumping, the technical and organizational innovativeness of farmers and farm-related agencies could be mobilized to put similar approaches into practice.

Research on the development of water and energy saving methods would enhance their applicability and effectiveness. The potential for research to improve our ability to obtain energy from alternative sources is indicated by the fact that the solar energy necessary to cause the evapotranspiration of one unit of irrigation water, is an amount capable of lifting that same amount of water 800,000 feet. Research on water and energy conservation and more efficient utilization should, therefore, continue to have high potential benefits to agriculture and society.

REFERENCES

Adams, R.M. "A Quadratic Programming Approach to the Production of California Field and Vegetable Crops, Emphasizing Land, Water and Energy Use." Unpublished Ph.D. dissertation, Univ. of California, Davis, 346 pp., 1975.

Adams, R.M., G.A. King, and W.E. Johnston. "Effects of Energy Cost Increases and Regional Allocation Policies on Agricultural Production." Amer. J. of Agr. Econ., 59(3): 444-455, 1977.

Chen, K.L. and R.B. Wensink. "Optimizing Oregon Agricultural Production Systems with Limited Energy Supplies." Transactions of the ASAE, 20(5): 986-991, 1977.

Davenport, D.C., R.M. Hagen, H.J. Vaux, and S. Hatchett. "Incidental Effects of Agricultural Water Conservation." Technical Completion Report (DWR Agreement No. B-53472). California State Department of Water Resources, Office of Water Conservation, Resources Agency, Sacramento, p. 93-116, June 1981.

Department of Water Resources. "The 1976-1977 California Drought: A Review." The Resources Agency, State of California, p. 50-72, May 1978.

Department of Water Resources. "Measuring Economic Impacts: the Application of Input-Output Analysis to California Water Resources Problems." Bulletin 210, The Resources Agency, State of California, p. 21-30, March 1980.

Dvoskin, D. and E.O. Heady. "U.S. Agricultural Production Under Limited Energy Supplies, High Energy Prices and Expanding Agricultural Exports." CARD Report 69, the Center for Agricultural and Rural Development, Iowa State Univ., Ames, 163 pp., November 1976.

Piper, D.B. "Drought of 1976-77 Central Valley, California." ERS Staff Report No. AGES82-624, NRED/ERS, U.S. Dept. Agr., Washington, D.C., 81 pp., July 1982.

Roberts, E.B. and R.M. Hagan. "Energy Requirements of Alternatives in Water Supply, Use, and Conservation: A Preliminary Report." California Water Resources Center, Contribution No. 155, Univ. of California, Davis, 113 pp., December 1975.

Williams, D.W. "Systems Analysis of the Flow of Energy, Nutrients and Water Through the California Agricultural Ecosystem." Unpublished thesis, Univ. of California, Davis, 180 pp., 1973

Williams, D.W. and W. J. Chancellor. "Irrigated Agricultural Production Response to Constraints in Energy-Related Inputs." Transactions of the ASAE, 18(3): 459-466, 1975.

12
Energy and the Limited Water Resource: Competition and Conservation

Joel R. Hamilton
Norman K. Whittlesey

INTRODUCTION

Other chapters in this book have viewed water as an input needed by crops for growth, and energy as a major input needed to supply the water. For some areas of the West this is an inadequate view of the interactions of water, energy and agriculture. While agriculture is a major user of water throughout the West, other claims on the limited water resources are made by municipal and industrial use, fish and recreation, waste disposal, transportation, and hydropower. Since nearly all water in the West is now being used for one or more of these purposes, the expansion of any one use often results in the reduction of some other use. This competition, especially the competition between hydropower production and irrigation, is the major subject of this chapter.

THE TRADEOFF BETWEEN IRRIGATION AND HYDROPOWER

The Historic Setting

Western irrigated agriculture traces back to the middle of the nineteenth century. The earliest irrigation invariably was located very close to the streams which served as water sources. The water could be diverted from the river by a low diversion dam, conveyed to the fields by gravity canals, and applied to the crops by gravity application methods. These techniques, requiring almost no energy except for human and animal power, allowed the development of irrigation on large acreages of land in a number of western states.

Later, as irrigation technology advanced, larger diversion and storage structures on the major rivers began to supply water to large irrigation projects. The improved diversion and storage dams, along with larger, better designed canals, allowed water to be supplied to lands some distance from the rivers. Still, through the first quarter of the twentieth century, most western irrigation required only gravity to divert, transport, and apply water.

Irrigation was, however, not the only use being made of the water. Use of water power to generate electricity began in the last decades of the nineteenth century. Many of the earliest hydroelectric installations were at natural falls along rivers, such as Snoqualmie Falls in Washington and Swan Falls in Idaho. As the market for electricity expanded, and the engineers' confidence increased, dams were placed across increasingly larger gorges. For example, American Falls Dam, built in 1913 on the Snake River in southern Idaho developed 49 feet of head. By the 1930s dams such as Grand Coulee, which developed 336 feet of head on the Columbia River, and Hoover, which developed 726 feet of head on the Colorado River, were being built to generate electricity.

Many of the early dams served both hydropower and irrigation purposes. They generated power and they either stored or diverted water for irrigation. The water might be released (generating electricity) and then diverted and used by downstream gravity irrigation systems. Other dams (such as American Falls Dam on the Snake River) served not only the storage and generation functions, but also elevated the water surface enough so that large new acreages could be served by gravity systems directly from the reservoir. Thus, even in the early stages of the dam building era irrigation had not yet become an energy intensive activity. In most parts of the West there was still ample water, both to generate the electricity that was needed and to provide for the crops being irrigated. Instream water use such as fisheries were being adversely affected but received little public or political attention until many years later.

By the 1930s and 1940s, most of the acreage easily accessible by gravity systems was exhausted and the technology of electric motors and pumps had improved, making pump irrigation a more attractive option. The Columbia Basin Project in central Washington began in 1952 to receive water pumped over 300 feet up from Lake Roosevelt formed by Grand Coulee Dam. Also, starting in the late 1940s, deep well pumping for irrigation became a significant trend in the West. At this point, in some parts of the West, irrigation and hydropower assumed a new competitive relationship. Irrigation had become a major consumer of energy for pumping, and that energy was supplied (in areas with abundant water and suitable topography) as electricity generated by falling water. This marriage between irrigation and hydropower was nowhere more apparent than in the Pacific Northwest. Because of the abundance of electricity that could be generated cheaply at hydropower dams along the Columbia River and its major tributary, the Snake River, it was natural that electric motors would power the pumps that served this region's energy intensive new irrigation. Much of the discussion in this chapter will focus on the Pacific Northwest because of the strength of this early symbiosis--now turned to competition--between irrigation and hydroelectric production. However, many elements of this relationship are also relevant to irrigation in other parts of the West.

In the Pacific Northwest the latest surge of new irrigation development started in the early 1960s and tailed off in the late 1970s. Much of the land first served during this period was located distant from, and on high benches above, the rivers. Pumping plants lifting water 500 or 600 feet to irrigate thousands of acres became common sights along the Columbia River in Washington and Oregon, and along the Snake river in Idaho and Washington. Starting in the 1940s, another trend had emerged--a shift from gravity application methods to sprinklers. While this change often allowed substantial improvements in irrigation efficiency and permitted rough and sandy lands to be irrigated, the change also greatly increased Northwest irrigated agriculture's use of electricity. Today, in the Pacific Northwest, irrigation has become a very significant consumer of electricity, both for pumping and for pressurizing sprinklers.

Competition Between Irrigation and Hydropower Generation

The abundant hydropower resource has been a boon to those parts of the West fortunate enough to have it. In 1974, hydropower accounted for 94 percent of the electricity generated in the three northwest states and as late as 1980 hydropower still provided over 80 percent of the electric power needs of the region (Wilkins). Most of the feasible dam sites on these rivers have already been developed. Development at most of the remaining technically feasible sites is blocked by political and environmental considerations. Consequently there can be little future growth in the Northwest's supply of hydroelectric power. When more electric generation capacity is needed to supply consumption requirements, the region must turn to alternative sources such as coal or nuclear generation. In fact the region is already heavily committed to this thermal alternative.

The region has, however, discovered in the last few years that the shift from hydro to thermal dependence is a painful and expensive process. The new thermal plants have proven much more expensive to build and operate than anticipated, and much more expensive than existing hydroelectric generating facilities. While existing hydro may produce power at costs ranging from a fraction of a cent to a bit over one cent per kwh, recently built thermal plants have current costs closer to four to six cents per kwh (Wilkins), and the thermal plants now under construction by the Washington Public Power Supply System could easily incur generating costs as high as seven to ten cents per kwh. As the proportion of the Northwest's electricity supply generated by thermal sources has increased in recent years, the rates paid by electric users has increased sharply to cover the higher costs.

Water resource development on the Snake-Columbia River system in Idaho, Oregon, and Washington has reached the point where nearly all of the water serves some beneficial purpose. Irrigation is the biggest consumptive use of water. Of the water that is not diverted and consumed for irrigation, the remainder is almost totally used at downstream hydropower dams. With the use of storage reservoirs to regulate water flows, only in rare spring floods does the Snake-Columbia flow exceed the capacity of the turbines, resulting in water flowing unused to the ocean.

Shortage of power for households and industry would cause enormous social and economic cost. Hence most electric power planning in the Pacific Northwest is based on "critical flows"--the lowest recorded historical flows over a consecutive 42 month period. So, while the region's hydropower plants frequently generate abundant surplus power for sale at low cost to agriculture, the aluminum industry, and export from the region, any changes in irrigation withdrawal do affect the generating potential of the region's rivers during drought. Any changes in consumptive water use by irrigation directly impact the firm electricity yield of the Columbia River hydropower system. Irrigation development imposes a significant cost on regional power consumers by indirectly causing all electricity costs to be increased as non-hydropower sources of electricity are added to the system.

The relation between irrigation and hydrogeneration in many other parts of the West is also competitive, although usually less so than in the three northwest states. For example, changes in irrigation water use in the Upper Colorado River Basin would affect generation at downstream dams, although the current level of water use in that region precludes much new irrigation. Except for the Pacific Northwest, and the Upper Colorado, most western irrigation is located below rather than above significant hydropower dams.

There are, however, a number of examples where the optimal operation of upstream water storage/power generation dams conflicts with the needs of downstream irrigation. This conflict is probably more typical of California and much of the Southwest. Much of the irrigation in the Southwest relies on upstream storage on the Colorado and the tributaries of the Sacramento (Norgaard). If water must be released to satisfy the quantity and timing requirements of irrigation, rather than optimally meeting the demands for electricity, then the hydropower value of the released water may be reduced.

The Importance of Other Water Uses. The introduction to this chapter mentioned a number of other uses of water--fish and recreation, municipal and industrial, and transportation. How do they fit into the picture? The final report of the Northwest Agricultural Development Project tabulated some of the uses of the Columbia River. It was estimated to have:

1. transported 7.5 million tons of freight (1.9 billion ton-miles of service) in 1979,
2. generated 126.4 million megawatt-hours of hydroelectric power in 1979,
3. provided 17.3 million acre-feet of water to irrigate cropland in Oregon, Washington, and Idaho during the 1975 crop year,

4. produced commercial fisheries output with an estimated market value of $20.1 million in 1978.

In the Pacific Northwest, current fishery management sometimes requires that water be intentionally released from dams to facilitate the passage of juvenile salmon and steelhead on their migration to the ocean. Such a conscious shift in the timing of Snake and Columbia River flows increases generation during the "fish flush" period of early summer and reduces power generation potential of the system during critically low water periods and higher power demand periods.

Other recreation uses also place constraints on the operation of the river. Minimum flows have been imposed on the Hells Canyon portion of the Snake River primarily to preserve float and jet boat recreation. Minimum flow regulations on the Grand Canyon portion of the Colorado and on streams such as the Merced in California also serve recreation boating needs.

In the Northwest, municipal and industrial water uses require only a small portion of the available water. In other parts of the West, these uses are the main competitors with irrigation for the limited water supplies. While the municipal water needs of the burgeoning Southwest population present an obvious conflict, the water requirements of energy developments such as shale oil and coal gassification in the Upper Colorado Basin represent a less obvious, but potentially very serious conflict (Department of Interior). Columbia and Snake River water transportation also requires minimum stream flows for barges, and uses water for lockage, so it too is a competitor for water, and places constraints on the operation of the river system.

Given this range of water uses, some of which consume water and others which place constraints on how the streams are regulated, what is the impact on the competitive relation between irrigation and hydroelectric generation? Society obviously places a very high value on fish runs, recreation, aesthetics, and on water for household use and waste disposal. Water use for industry and transportation has high economic value. In the West, the increasing pressure of these growing water uses can only intensify the competition between water use for irrigation and water for what seems to be the residual use--hydropower generation (Wharton and Whittlesey).

Irrigation Versus Hydropower in the Pacific Northwest. When a water resource is fully used for various productive purposes an expansion of any one use means that some other uses must be reduced. Most important in the context of this chapter, there is a direct competition between irrigation and hydroelectric generation for the use of Snake and Columbia River water. Any expansion in consumptive use by irrigation must result in less water available at all downstream points for power generation.

Table 12.1 shows estimates of the hydropower loss when water is diverted and consumptively used at various points in the Snake-Columbia River system. For example, water from American Falls Reservoir in southeast Idaho could potentially be passed through the power plants of 21 existing sets of turbines on its way to the Pacific Ocean (Hamilton and Whittlesey). Of the 4,297 elevation foot drop from the American Falls Reservoir surface to sea level, about half (2,094 feet) has been developed for power generation. An acre-foot of water dropped through one foot of head generates about .87 kwh of electricity. Thus an acre-foot of water released from American Falls Reservoir could potentially generate 1,821 kwh of electric power if it passed through each of the 21 power plants. Water from the reservoir above Grand Coulee Dam on the Columbia River in Washington could be used at 11 existing downstream power plants where is could generate 1,015 kwh of electricity per acre-foot. One consequence of irrigation diversion is, thus, to take water away from the hydroelectric system, reducing hydropower generation, and forcing the region into even greater dependence on expensive thermal generation.

However the loss of hydropower is only part of the impact of irrigation. As noted above, most recent irrigation development has been very energy intensive. Most recent development has tended either to use deep wells or be located on high benchland above the rivers which serve as the water source. Thus, large quantities of electricity are consumed for pumping. In addition most such new irrigation uses sprinklers, so electricity is needed for system pressurization. Electricity use depends on the water use efficiency, lift height, friction losses, and the operating pressure of the irrigation system. In approximate terms, these systems require 100 to 200 feet of head for sprinkler pressurization and the power use is about 1.25 kwh per acre-foot of lift and pressurization (Hamilton, Barranco and Walker). For example, irrigation with 500 feet of lift plus 200 feet for pressurization

Table 12.1.
Potential energy lost by consumptively diverting an acre-foot of water from the Snake-Columbia system

Item	Cumulative Head (feet)	Cumulative Energy at .87 kwh/acre ft/ft (kwh)	Cumulative Value at 50 mills/kwh (dollars)
Columbia River (Washington-Oregon)			
Bonneville	59	51	2.55
The Dalles	142	124	6.20
John Day	242	211	10.55
McNary	316	275	13.75
Columbia River (Washington)			
Priest Rapids	393	342	17.10
Wanapum	470	409	20.45
Rock Island	504	439	21.95
Rocky Reach	591	514	25.70
Wells	658	573	28.65
Chief Joseph	825	718	35.90
Grand Coulee	1167	1015	50.75
Snake River (Washington)			
Ice Harbor	414	360	18.00
Lower Monumental	514	447	22.35
Little Goose	612	532	26.60
Lower Granite	710	618	30.90
Snake River (Idaho-Oregon)			
Hells Canyon	920	800	40.00
Oxbow	1040	905	45.25
Brownlee	1312	1141	57.05
Snake River (Idaho)			
Swan Falls	1336	1162	58.10
C.J. Strike	1424	1239	61.95
Bliss	1494	1300	65.00
Lower Salmon Falls	1553	1351	67.55
Upper Salmon Falls "A"	1599	1391	69.55
Upper Salmon Falls "B"	1636	1423	71.15
Shoshone Falls	1850	1610	80.50
Twin Falls	1997	1737	86.85
Minidoka	2045	1779	88.95
American Falls	2094	1821	91.05

Source: Hamilton and Whittlesey.

would require about 875 kwh to pump each acre-foot of water applied to a crop. If 3.5 acre-feet of water is used per acre, then 3,063 kwh of pumping electricity is used for each acre irrigated.

The total electrical energy impact of recently developed energy intensive high lift irrigation in the Pacific Northwest can easily be as much as 6,000 kwh per acre per year--3,000 kwh used for pumping and 3,000 kwh of lost hydropower. The actual figure depends on location (how many dams there are downstream and how many dams are bypassed by any return flows), on lift height, and on the quantity of water use by the crops grown. Specifying the cost of generating thermal electricity to serve such irrigation and replace the hydro loss is subject to some debate. About five cents per kwh is a rather arbitrary estimate of what it currently costs to operate a thermal powerplant. At this rate, it costs about $300 per year to generate the 6,000 kwh of electricity used and hydropower lost to a typical acre of energy intensive high lift irrigation in the Pacific Northwest.

Of course, the rates paid by irrigators are not these high rates reflecting the marginal cost of new electricity generation. Such costs would be prohibitive to both existing and new irrigation. The rates which farmers actually pay for pumping energy reflect both preference status and average cost pricing. Since there is still a large amount of cheap hydropower being blended with the growing amounts of expensive thermal power, the average cost being charged to northwest farmers is still far below the marginal generating cost, and below the power rates being charged elsewhere in the country. The preference status is partly unofficial (the tendency of rate setting agencies to award lower rates to agriculture) and partly official (a preference status that the Northwest Power Bill has now extended to residential and farm users in the region).

In 1980, the amount of a farmer's pumping power bill which actually went to cover costs of electricity generation, after deducting the costs of transmission, administration, and billing averaged .43 cents per kwh in the Columbia Basin portions of Oregon and Washington, and probably about .7 cents per kwh in Idaho (Whittlesey, et al., 1981a). Farmers in Bureau of Reclamation projects such as those in the Columbia Basin of Washington receive more favorable treatment, since the rate such projects are charged for electricity used to deliver water to farms was set many years ago by contract at .05 cents per kwh.

Farmers involved in new irrigation development, of course, pay nothing to defray the cost of replacing the lost hydroelectricity. They pay only average costs for the pumping electricity used. What they do pay is often substantial and sometimes financially devastating. However the water itself, is free. Thus the difference between the marginal cost of generating the replacement power, and the payment which farmers make for the power they use is an externality or subsidy paid by all consumers of electricity in the Pacific Northwest in the form of higher electricity rates. An acre of new irrigation development could impose a total energy cost of $300 per acre, as in the above example. If the farmer pays three cents per kwh for electricity used, his power bill is $90 per acre. The balance, $210 per acre, becomes a perpetual subsidy paid by other consumers of electricity.

The Role of Groundwater. It is obvious that diversion of streamflow affects hydropower generation. How does groundwater pumping fit into this picture? In most of the Pacific Northwest there is a very close linkage between groundwater and streamflow. The Snake River Plain aquifer in southern Idaho is fed in some places by streams that disappear into the lava beds, only to reappear later as the groundwater spills back into the Snake River. Similar, if less spectacular, linkages between river and aquifer occur in other parts of the West. Since water movement through these aquifers is often very slow, the effect of groundwater pumping for irrigation may not be reflected in reduced streamflow until months or years later. But groundwater pumping in such areas will eventually reduce hydropower production and have the same kind of energy impacts as described for surface water irrigation development. In other areas of the West, groundwater pumping does not have any appreciable hydropower impact. In some areas this is the result of the relative isolation of the aquifer (such as the Ogallala in the Southern Plains states, the Odessa in Washington, and the Raft River Aquifer in Idaho) from streamflow. In other areas such as the Central Valley of California, the obvious reason why pumping from wells does not effect power generation is because there are no hydropower dams downstream from the natural out-flow of the aquifer. Except for these regions where the aquifer is truly isolated from streamflow and where groundwater mining is possible (although at the expense of increasing pumping costs), changes in water use by deep well irrigation can ultimately affect hydropower generation just as surely as irrigation from surface sources.

The Battle Over New Irrigation Development

Most regions of the West are not concerned with the potential impacts of large new irrigation developments since they have neither the required arable land nor the required water. Only in the Pacific Northwest and parts of the Great Plains are the land and water resources physically available to support such development. In the Pacific Northwest, however, new irrigation development has become the focus of emotion charged controversy. The following is a discussion of the economic and political forces at work stimulating and restricting irrigation development in the Northwest states.

The Economic, Political and Institutional Setting. The political power and popularity of irrigation development in the West and particularly the Pacific Northwest still runs at a very high level. Few public officials at any level of government are willing to question the wisdom of further irrigation development. Oregon, Washington, and Idaho each have official plans for further development of new irrigation. These plans call for approximately 2.2. million acres of additional irrigation in the three states by year 2000. The recent economic conditions and the slowing rate of irrigation development have tempered these estimates of likely new irrigation. However, the public popularity and the political power of the irrigated sector remain high and are expected to do so for many years to come. If opportunities exist and the economy permits further development, there will certainly be ample public support for that activity.

As further stimuli for irrigation development, there are several subsidies available from state and federal agencies. The most obvious of these agencies is the U.S. Bureau of Reclamation. The completion of the Columbia Basin Project in central Washington remains a high priority for the USBR and for many of the local leaders in that region. In such projects the capital subsidies for the development activity are very high, and since funds are provided by the federal government, such subsidized irrigation development would certainly stimulate regional economic growth.

Other examples of development subsidies include the Carey Act and the Desert Land Act, which could potentially stimulate new Irrigation in Idaho by making federal land available very cheaply to would-be developers. A recent Bureau of Land Management Environmental Impact Statement mentioned the existence of the irrigation--hydropower tradeoff (BLM). However, the BLM controlled public lands in Idaho are still open for possible development, and any decision to actually release these lands will still be made by BLM based primarily on private development feasibility, with little if any recognition of the regional energy cost impacts.

The states have also encouraged new irrigation by financially participating in the development. Idaho, Oregon, and Washington each have provided some means of subsidizing irrigation growth. These programs generally have the effect of providing state funded interest subsidies for capital investment in new irrigation.

Low electricity rates, low interest rates, and relatively favorable prices stimulated irrigation from the late 1960s through most of the 1970s. All of these factors have changed dramatically now and the economic feasibility of irrigation development appears to have declined significantly.

Leading the list of inhibiting factors are the rising electric rates. Pacific Northwest electric rates seem likely to continue to increase, in real terms, through at least the 1990s. The agencies administering and regulating electricity supply and pricing are now becoming aware that new irrigation is a growing new load. However, these agencies face both legal and political barriers to implementing more creative pricing systems, and constitutional barriers against refusing to serve any class of new users such as new irrigation load.

Irrigation development is a very capital intensive activity. Most new developments in the Northwest today incur capital costs ranging from $1500 to $2000 per acre. The high lift and distance from water sources to potentially developable lands is also a deterrent to growth. Those lands which remain undeveloped will require high lifts and long pumping distances in order to get water to the land. These factors translate into high investment and operating costs.

An Idaho study (Hamilton, Barranco, and Walker), has examined the sensitivity of new development to changes in factors such as electric rates. The study used a computer budgeting model to simulate the investment costs and estimate electricity costs of irrigating new land at various distances from, and heights above the Snake River water source. The result was an estimate of how high and how far it is privately feasible to pump irrigation water. These results were applied to actual lifts and distances for an 111,000 acre tract in southwest Idaho which the Bureau of Land Management is considering irrigation development. Implicit in the procedure was the assumption that the relative relationship of crop prices to production costs exclusive of electricity would remain as it was from 1974 to 1977, a period that in retrospect was one of rather favorable costs and prices. The results, for three different possible farm sizes, is given in Figure 12.1. The figure shows the estimated acreage within the 111,000 acre tract which could feasibly be irrigated at various percentage electric rate increases above the 1978 rates used as a base. Note that real electricity prices in this region have risen 200 to 400 percent since 1978, essentially eliminating irrigation development in many areas. The extreme sensitivity of new irrigation feasibility to electric rate increases is a phenomenon that applies across the entire West.

Because of the interest in and concern over the impacts of new irrigation, its progress is watched with interest. In the decade prior to 1978, Northwest land was being developed for irrigation at an average rate of about 100,000 acres per year. Of this, Idaho developed about 50,000 acres per year and Washington and Oregon each approximately 25,000 acres per year. The more recent rate of development has been far slower.

Lower estimates of likely development in Idaho are reinforced by a recent study which concluded that even if the legal problems restricting Idaho irrigation are resolved, new development by 2000 will probably total less than 195,000 acres (Hamilton and Lyman).

It is at least possible that growing awareness of the energy tradeoffs associated with new irrigation may begin a feedback process, stimulating formation of new policies and institutions for regulating new development. A unique situation is now developing in Idaho, where state courts have recently held that a hydropower dam built on the Snake River at the turn of the century holds water rights superior to upstream irrigation. This situation has for several years been an effective block to new irrigation in much of southern Idaho, and even jeopardizes the water

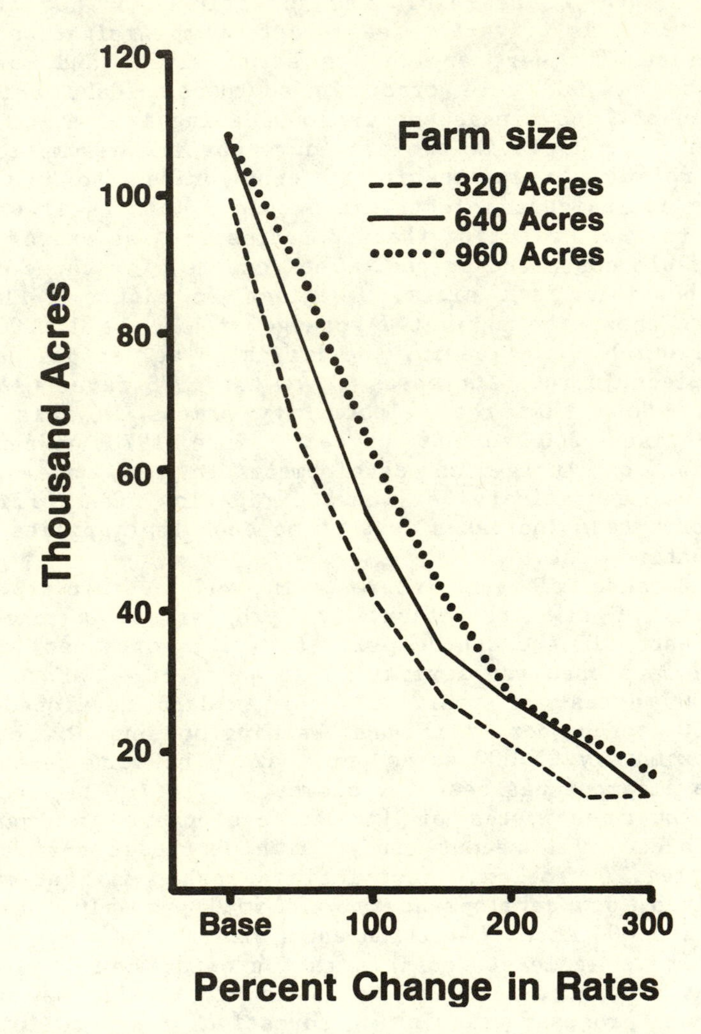

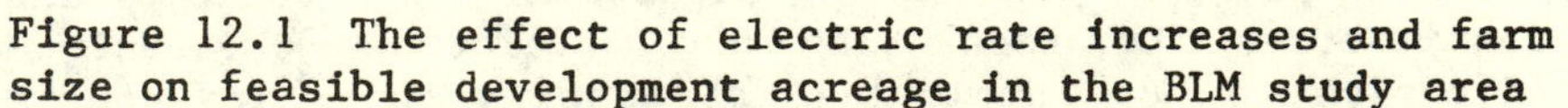
Figure 12.1 The effect of electric rate increases and farm size on feasible development acreage in the BLM study area

supplies of some recently developed lands. While it is likely that legislative action will protect existing irrigated acreage, the case has very graphically presented southern Idaho residents with the reality of the finite water resource, and the real possibility that the opportunities for further irrigation development may be very limited.

Other Regional Development and Social Infrastructure Impacts. In assessing the impacts and desirability of new irrigation it is important to view this development in the context of the local community and the wider economy. Both the energy and the economic impacts of irrigation development extend far beyond the acres being irrigated. The residences of the farmer and farm laborers consume electricity. If the irrigated production is processed, and much of it is, this also consumes electricity. The farmer earns income from his irrigation activities, and spends that income on both production inputs and personal consumption. This generates more income and more jobs in the local and regional economies, and results in further increases in the demand for electricity.

It becomes difficult even to determine which are costs and which are benefits. Irrigation development, because it forces the replacement of hydroelectric generation with thermal generation, creates jobs and income from powerplant construction and operation. Is this a benefit to the region or a cost because a higher cost technology must be used to produce a given product? No one has yet developed an adequate methodology to sort out these various impacts of new irrigation, but there have been several attempts at partial analysis. Hamilton and Whittlesey estimated that 1 on-farm and 1.8 off-farm jobs are created by each 100 acres of new irrigation. They also estimated that the annual social cost of each new job (the cost of generating the power to replace the lost hydropower and the electricity consumed by irrigation pumping, residences, processing, etc., plus the cost of other local infrastructure such as schools, roads, and other government services for the new people) totaled as much as $10,000 per job per year in 1978.

Two studies have attempted to cast the irrigation development in an input-output analysis framework (Hamilton and Pongtanakorn, and Findeis and Whittlesey). Both studies, one focusing on southern Idaho, and the other on the Columbia Basin Project of Washington, demonstrated that irrigation development generates large amounts of regional income, and many jobs. It is not clear from these studies whether regional income gains may exceed the losses from the electricity cost increases. What is important however, is that the gains and losses are not distributed uniformly across the sectors of the economy, with households and energy intensive sectors such as the aluminum industry and present high lift irrigation being major victims of the electricity cost increases caused by development. Most importantly, the Washington study showed that no other economic sector benefitted enough from new irrigation that it would be likely or even able to subsidize the development. This result demonstrates why future Northwest irrigation development will depend on the various subsidies available from the federal and state levels.

Another input-output based study, done by the Bonneville Power Administration looked at the question from a somewhat different perspective (Wilkins). The question asked was what loss would result if the electric power supplies of various industrial sectors were restricted. Table 12.2 shows estimates of direct and indirect output and wage loss resulting from a hypothetical power deficit in the Pacific Northwest. The position of pump irrigated crops in this table has implications not only for the desirability of new irrigation, but also for the position of energy intensive irrigation in energy shortage contingency planning. If cheap hydroelectricity is considered a scarce and limited resource, then economic development would be most efficient if it avoids electricity intensive industries. Shifting emphasis away from pump irrigation toward less electricity intensive industries would preserve the Northwest's hydroelectric resource and might be a more efficient way to create jobs and income for the region.

IMPLICATIONS FOR ENERGY AND WATER CONSERVATION PROGRAMS

This chapter has outlined the competitive relationship between irrigation and hydropower generation, and noted some of the controversy regarding the desirability of developing new irrigation. This interaction between irrigation, energy use, and energy production also has implications for energy and water conservation programs.

The Design and Effectiveness of Water and Energy Conservation Programs

Both energy and water conservation have received a lot of recent attention. Programs have been initiated at the state and federal level to improve on-farm irrigation water and energy efficiency. Electric utilities have also developed energy conservation programs. Higher energy costs and narrower profit margins have stimulated farmer interest in techniques and technologies that would conserve water and energy. What is the relationship between water and electricity conservation programs and by the competitive relation between irrigation and hydropower generation?

The important conclusion is that improving water use efficiency is sometimes a good way of achieving electricity conservation--but not always. For example, improving the application efficiency of a high lift irrigation project results in less electricity being needed to pump a reduced quantity of water. But, in spite of the reduced diversions, there may be an increase in hydropower loss attributable to the project if the improved application efficiency reduces irrigation return flows. Often improvements in water application efficiency are achieved by energy

Table 12.2
Estimated wages and output lost due to hypothetical electricity shortage

Industries	Dollar Loss per kwh not Delivered in 1985 (dollars)
Food Processing	10.90
Lumber and Wood	5.30
Pulp and Paper	.90
Chemicals	.66
Pump Irrigated Crops	.48
Aluminum	.07

Source: Wilkens.

intensive means such as conversion from gravity application to sprinklers. This may actually increase water consumptive use, while adding a new electricity load.

If water consumption by irrigated crops can be reduced, then more water will be available for generating power. However, reducing water consumption, without simply reducing acreage is a difficult task. Changing crops, deficit irrigation, reducing evaporation during storage and delivery, and reducing water loss to phreatophytes (canal-side plants, brush, and trees) are measures that can conserve water for hydroelectric generation.

Since a large part of the tradeoff between irrigation and hydropower is in the form of an externality borne by all users of electricity, society has much more reason than irrigators, to be interested in effective water and energy conservation programs. The costs faced by the farmer induce him to apply less emphasis on water conservation than may be socially optimal, and allow him to ignore some of the energy interactions among conservation practices. Programs by state and federal agencies, and public utilities to encourage, subsidize, and guide electricity and water conservation activities may be justified on these grounds.

Contingency Planning for Water and Energy Shortages

One very important role not now being effectively filled by government or any other institution is the task of contingency planning for possible energy or water shortage. What emergency conservation programs should be developed to deal with short term shortages of energy or water? At issue is the treatment that should be given to irrigated agriculture if the region is faced by drought caused shortages, or perhaps by renewed problems in the Middle East. Where should irrigation be in the priority ranking of electricity and water uses? In the 1970s when concern over energy shortage was a public issue, it was determined that agriculture was to receive high priority in using available energy supplies because food is important (and because agriculture is a powerful political lobby). In hindsight, at least for the Pacific Northwest, that answer looks simplistic.

Most western water law is very clear regarding the drought-time priority of irrigation diversions (Hamilton,

Walker, Grant, and Patterson). In midsummer, after crops are planted and growing, depriving them of water to save energy might cause severe economic damage relative to the energy saved. However, a decision in the spring not to plant some high lift acreage could save energy at far less economic cost. In fact, it might be possible through contractual agreements, rate design or other institutional changes to induce (or force) irrigators to shift water and energy use away from low flow years or portions of the year (Harrer and Wilfert). While such programs might be costly to farmers, these costs could perhaps be offset by subsidies or other inducements, which might leave the region better off by increasing firm power generation and delaying the need for building thermal powerplants.

While energy shortage is not now an immediate concern, a lesson that should have been learned from the last decade is that it would be prudent for agencies to give some thought to energy contingency planning, to irrigation's role in such plans, and to the possibility of changing the rules governing water use to allow more flexible response to such emergencies.

SUMMARY

This chapter has examined the implications of the competition between irrigation and hydropower for the limited water resource of the West. The principal conclusion of the chapter is that it is essential to consider the whole range of potential impacts when considering policies and programs related to irrigation and energy use in irrigation. Because of the intimate linkage between irrigation and hydroelectric generation in the Pacific Northwest, irrigation decisions must consider energy issues, and energy decisions must consider irrigation. Decision makers and regulatory institutions must maintain this wider perspective when dealing with issues such as irrigation energy and water conservation, energy and water shortage contingency planning, and the allowable level and rules governing future irrigation development.

REFERENCES

Eckstein, O. "Will Higher Interest Rates Kill the Recovery?" U.S. Forecast Summary, Data Resources Inc., May 1983.

Findeis, J. and N. Whittlesey. "Competition Between Irrigation and Hydropower Water Use in Washington State." State of Washington Water Resource Center Report 44, Pullman, June 1982.

Hamilton, J. and N. Whittlesey. "Social Costs and Energy Impacts of Irrigation Expansion in the Pacific Northwest." Paper presented to the Pacific Northwest Conference on Nonfederal Financing of Water Resource Development, Portland, OR, January 1978.

Hamilton, J., G. Barranco, and D. Walker. "The Effect of Electricity Prices, Lift, and Distance on Irrigation Development in Idaho. " Amer. J. of Agr. Econ., May 1982.

Hamilton, J. and C. Pongtanakorn. "The Economic Impact of Irrigation Development in Idaho: An Application of Marginal Input-Output Methods." Annals of Regional Science, forthcoming.

Harrer, B. and G. Wilfert. "An Assessment of the Use of Interruptible Electric Power in Irrigated Agriculture." Battelle Memorial Institute Pacific Northwest Laboratories, Richland, WA, January 1983.

Norgaard, R. "Economic Analysis of Additional Hydropower Development on the Tuolumne River." California Dept. of Water Resources, Sacramento, November 1982.

Northwest Agricultural Development Project. "Final

Report." PNW Regional Commission, Vancouver, WA, June 1981.
Northwest Power Planning Council. "Columbia River Basin Fish and Wildlife Program." Portland, OR, November 1982.
Wharton, R. and N. Whittlesey. "The Hydropower Cost of Irrigation Development and Fisheries Enhancement in the Columbia River Basin." Paper presented at the PNW Regional Economic Conference, Boise, ID, April 1981.
Whittlesey, N., et al. "Energy Tradeoffs and Economic Feasibility of Irrigation Development in the Pacific Northwest." Agr. Res. Bull. 0896, Washington State Univ., Pullman, 1981a.
Whittlesey, N., et al. "Demand Response to Increasing Electricity Prices by Pacific Northwest Irrigated Agriculture." Agr. Res. Bull. 0897, Washington State Univ., Pullman, 1981b.
Whittlesey, N., et al. "Demand for Electricity by Pacific Northwest Irrigated Agriculture." Consultants report to Bonneville Power Administration, January 1982.
Whittlesey, N. "Energy and Irrigated Agriculture: Problems and Prospects." Paper presented at Agricultural Conference Days, Corvallis, OR, March 1983.
Wilkins, J. "Economic Issues in Competition for Columbia River Water - Hydropower." In Economic Issues in the Competition for Columbia River Basin Water, PNW River Basins Commission, Vancouver, WA, March 1981.
U.S. Department of the Interior. "Report on Water for Energy in the Upper Colorado River Basin." July 1974.
U.S. Bureau of Land Management. "Boise District Agricultural Statement for Southwest Idaho." Boise District Office, 1979.

13
Production of Livestock with Irrigated Lands

Gerald M. Ward
John F. Yanagida

HISTORICAL DEVELOPMENTS

The arid West, in its original condition, was a very fragile environment for agriculture. The early days of open range grazing were fraught with struggles between ranchers and homesteaders as well as those between cattle and sheep grazers. Drought, occasional severe winters, and uncertain feed supplies made livestock production a high risk venture. In addition, the tremendous distances from the markets made for returns that were only marginal. The development of irrigated lands made possible an integration of livestock and crop production which ultimately produced a flourishing livestock industry in the West.

The arid grazing lands of the vast western area were adequate to support large herds of cattle and flocks of sheep for most of the year. However, the hazards of drought and severe winter weather required some supplemental feeds. With the development of irrigation, supplemental feeds could be produced in the proximity of grazing livestock herds. Relatively small amounts of supplemental feeds can have a pronounced effect upon the productivity and output of herds. The lack of this input is a major constraint on grazing animal production throughout much of the world today.

Irrigation had an even greater impact on the output from livestock enterprises through the development of the feedlot. Removal of young animals from the herds for finishing to slaughter weights by more intensive feeding systems has had a tremendous effect on range livestock productivity. Instead of maintaining a steer on grass for four or five years before slaughter, the animals now could be marketed at two years of age. The result is that the grazing resources can support much larger breeding herds. Feeding systems were developed originally in the West to exploit use of alfalfa hay, a prime cash crop produced under irrigation. Early in many areas of the West, the sugar beet industry also developed on irrigated lands. Beet pulp provided an excellent feed for finishing cattle and sheep. Fort Collins, Colorado, for instance, became the largest lamb feeding area in the country based primarily upon alfalfa hay and beet pulp in earlier days. Lambs were procured from the huge flocks that wintered in the mountain valleys or plains and spent the summer grazing in the high mountain areas of Colorado. Cattle feedlots developed somewhat later and were an even larger consumer of products from irrigated land. Some of the largest feedlots in the world were developed in northern Colorado and utilized corn grain, corn silage, and alfalfa hay for finishing large numbers of cattle. A later development in cattle feeding occurred as underground water supplies were exploited. This occurred first in the Panhandle of Texas and later spread north to western Kansas and Nebraska. Exploitation of this water supply altered drastically the beef cattle production picture for the United States as the feedlots and packing plants moved to these areas from the original area of farmer feeders located in the midwest Corn Belt.

As irrigation developed, dairying became a substantial part of general farming on irrigated farms in the West. The demand for fluid milk was very limited in this area of low population density, but there was a market for cream or milk for cheese factories in most of the irrigated areas. Dairy production existed mostly on irrigated lands with irrigated pasture and combinations of silage, hay and grain. Dairy breeding stock might be pastured on range along with beef cattle, but milk production had to be supported primarily from crops grown on irrigated land. As population grew in the West, the rapid development of a "feedlot-type" dairy industry also developed rapidly.

The relationship between range grazing areas and irrigated crop production is perhaps illustrated best by the relatively small amount of swine and poultry production in western irrigated areas. Poultry production is prominent in California but feed resources from irrigated land have primarily been used to increase the value of grazing animals or the output from grazing animal systems. In summary, the contribution of irrigation to western livestock production has been the production of supplemental feeds. These feeds enabled the development of large scale feedlots, dairy operations, and larger breeding herds. This resulted in upgrading the quality and value of associated livestock operations.

ENERGY INPUTS FOR FEED PRODUCTION

Energy requirements for livestock production are largely embodied in feed production. The feed component comprises over 50 percent of the total energy used in livestock production. Energy use per unit of feed varies from a low input for unmanaged grazing lands to high levels for grain or forage produced with deep well irrigation systems.

Energy analyses of crop production have been published for a variety of crops and regions in the years since 1973. The most comprehensive was prepared by the Federal Energy Administration and the U.S. Department of Agriculture which presents data by farm operation and fuel type for each state.

Estimates of energy requirements for producing major feed crops by regions of the United States are presented in Table 13.1. Corn is the basic feed grain for most of the country. Estimates of energy requirements for the Corn Belt are about 1,800 BTU/lb of corn, and that region accounts for a high percentage of the nation's total production. Estimates for northeastern states are similar, but the Southeast is higher primarily because of lower yields with about the same energy inputs per acre. These three regions produce grain mostly without irrigation so that nitrogen fertilizer represents the largest single energy input. Corn production under irrigation in the western states, although providing higher yields, is more energy intensive if the irrigation water has to be pumped. Stewart estimated 3,000 BTU/lb for corn production in West Texas compared to 1,170 BTU/lb in Iowa.

Variation in energy requirement estimates result from assumptions about types of irrigation systems and sources of power. Surface irrigation from established reservoirs and canals requires little energy input except for manual labor, if the energy embodied in the construction of irrigation facilities is not considered. Whittlesey has shown that very energy intensive irrigation systems have

Table 13.1
Energy use in feed crop production BTU/lb x 10^3

	Dryland			Irrigated	
Region	Northeast	Southeast	Corn Belt	West	Colorado
Corn Grain	1.50-16.3	2.33-3.03	1.39-1.81	1.81-4.4	3.31[a] 1.35[b]
Sorghum Grain	1.48-2.15			1.73-5.67	
Wheat (Great Plains)		1.59-2.69			1.30[a]
Corn Silage	.99	1.12	1.08	1.6	1.75[b]
Hay (Timothy)	.45	1.10	.87	3.99[d]	1.30
Alfalfa Hay		.54		2.08 .89	1.16[1] .36[c]
Dehydrated Alfalfa			3.61	4.23 .87	5.00[a]
Soybeans		3.02	1.63		
Urea		26.50	26.50		

[a]Ward, 1980.
[b]Cervinka et al.
[c]Cook et al., 1976.
[d]Arizona data, Pimentel.
All other data from Pimentel.

developed in the Northwest in response to available cheap electricity generated from the vast hydroelectric network in that region. Dvoskin, in another study, concluded that increasing energy costs will reduce grain production from western irrigation with a concomitant increase in returns for Corn Belt farmers.

Sorghum is the major feed grain grown both on dry land and under irrigation in the hot, arid Southwest. Its energy requirements are essentially the same as for corn production (Pimentel). Other less important feed grains on a national scale but important regionally are barley, oats, and rye. Wheat is grown primarily as a food grain, but large amounts are often used for livestock feed. Wheat is produced primarily on the Great Plains without irrigation and with only limited use of fertilizer; however, energy inputs per unit of grain are about the same as dryland corn production (Pimentel; Ward, et al.).

Corn silage is an important feed source for ruminant animals and contains only about 30 percent dry matter. The energy estimates in Table 13.1 are based on dry matter weight to account for a lower energy input for silage than corn grain. The residue, corn stover, represents about one-half the dry matter of the plant. This may be grazed by cattle, harvested as feed or returned to the soil after harvesting the grains.

Hay is a term applied to a great variety of grasses and legumes. Management of the different varieties varies from minimal attention for old stands of natural grasses to intensive attention for seed-bed preparation, regular fertilization, pest control and irrigation for specialized alfalfa hay producers. Hay may be grown on the best crop land in rotation with field crops or it may be harvested from land unsuited for other crops because of flooding, erosion, or drought potential. Energy requirements for hay were calculated nationwide by Pimentel as summarized in Table 13.1. Alfalfa and other legume crops do not require nitrogen fertilizer to generally reduce the total energy input.

Dehydrated alfalfa is an important supplemental feed for both cattle and poultry, because it provides relatively high protein content, carotene as a precursor of vitamin A, and xanthophyll which produces the desired yellow color in eggs and broiler chickens. Immature high moisture alfalfa is chopped and dried with natural gas to produce dehydrated alfalfa. The energy requirements for drying are three to four times the energy for producing the crop.

Nearly all livestock enterprises require protein supplementation. A diverse group of products are used as protein supplements, but soybean meal is the most widely used, although not produced in the West. Calculation of energy used to produce soybean meal illustrates a common problem when dealing with byproducts, that is, how to allocate the energy input between two or more products (oil and meal). Since the value of oil and meal are roughly equal, although the products are 80 percent meal and 20 percent oil, we have divided energy inputs equally. Energy for production ranges from 1,600 to 6,600 BTU/lb of soybeans (Pimentel), and extracting energy is estimated at 1,250 to 4,300 BTU/lb (Ward, 1982) which means that soybean meal may require 2,850 to 10,800 BTU/lb. Cottonseed meal calculations are equally complex, because cotton lint is the primary product.

Urea, a nonprotein source, can be used by the microbes in the stomachs of ruminants to synthesize protein if the ruminant diet contains enough digestible energy which generally means a high percentage of grains. Production of urea requires 26,500 BTU/lb, and since urea supplies no feed energy, the general rule is that one pound of urea plus seven pounds of corn is equal to eight pounds of soybean meal. Thus, although economics may dictate substitution of urea for soybean meal, it will not improve energy efficiency.

ENERGY INPUTS FOR LIVESTOCK OPERATION

Estimating the energy requirements for producing livestock increases in complexity with diversity of management practices, intensification, and fragmentation of the enterprise. Complexity increases and intensification decreases in the following order: poultry, swine, dairy, sheep, horses, and beef cattle.

Poultry and swine are non-ruminants and generally do not compete with cattle, sheep, and horses for forage supplies. Both swine and poultry production have become highly specialized commercialized operations providing a uniform product as contrasted with beef and lamb. The very large confinement operations are largely dependent on purchased feed which may be produced distant from the feedlot. However, neither swine nor poultry operations are major enterprises in the West.

In terms of total energy use, beef cattle use substantially more energy than other species, followed by dairy cattle, swine, and broilers (Table 13.2). Energy per unit of production cost is lowest for laying hens and highest for turkeys while the differences among other species are probably not significant. Energy requirements are primarily determined by feed production, although swine and poultry require some heating of buildings. Table 13.3 presents estimates of energy use for egg and broiler production plus other livestock products.

Swine feeding and management systems vary by regions of the country, but Pimentel indicated only small differences in energy use by regions except for the Southwest where the feed consists primarily of irrigated sorghum

Table 13.2
Energy use in United States livestock production[1]

Type of Livestock	Total National Energy (10^{13} BTU)	Energy per $1 of Production Cost ($10^{3}$ BTU)
Cow-Calf	6.62	6.16
Beef Feedlot	2.17	
Dairy Cows	5.21	5.36
Hogs	3.72	5.40
Layers	.56	3.85
Broilers	2.00	8.23
Turkeys	.72	10.57
Sheep and Lambs	.37	6.84

[a]FEA-USDA Energy and U.S. Agriculture 1974 Base. Includes all energy used on farm for crop and livestock purposes. Total estimate for all livestock is 0.224×10^{15} BTU.

raising the energy intensity for producers in that region. Energy requirements per pound of liveweight gain are slightly less for pigs than for broilers, while carcass meat as a percentage of liveweight is about the same for pigs and broilers and considerably higher than ruminants (i.e., sheep and cattle).

Dairy production, especially in the West, is highly specialized. Dairy cows selected for very high milk production require high quality feeds. The amount of grain fed to dairy cattle has increased almost threefold over the past 25 years while milk production per cow doubled (Coppock, et al.). Energy requirements do not vary greatly among regions (Table 13.3), primarily because the more energy intensive production systems result in greater production per cow; this is particularly true for California and Arizona. Energy for feed ranges from 62 to 70 percent of the total energy use. Ward (1980) estimated energy usage for milk production on an irrigated farm producing corn silage and alfalfa and purchasing grain to be 2,400 BTU/lb of milk.

Table 13.3
Energy use in broilers, eggs, pork, lamb and milk production (BTU/lb)

Item	Energy per Unit Product (10^4) BTU/lb	Energy per Unit Protein (10^4) BTU/lb
Broilers		
Northern U.S.[a]	1.55	--
Southern U.S.[a]	1.30	--
U.S. Average[b]	--	3.99
U.S. Average[c]	1.26	--
Eggs		
Northern U.S.[a]	1.70	--
Southern U.S.[a]	1.66	--
U.S. Average[b]	--	2.37
U.S. Average[c]	1.45	--
Swine		
U.S.[a]	1.14-1.73	12.83-19.69
(Live weight gain)		
U.S. Average[b]	--	6.39
U.S. Average[c]	1.99	--
Lamb		
Midwest[a]	7.55	--
Western Range[a]	1.46	--
Western Range[d]	.56	--
(Live weight gain)		
U.S. Average[b]	--	2.93
U.S. Average[c]	1.97	--
Milk		
U.S. by Regions[a]	.21-.33	5.60-9.21
U.S. Average[b]	--	6.48
Colorado[c]	.24	--
U.S. Average[c]	.32	--

[a] Pimentel.
[b] Pimentel et al.
[c] Pierotti.
[d] Cook et al., 1976.
[e] Ward, 1980.

Energy estimates for dairy production usually consider milk as the only output, but beef and veal are important products on dairy farms. It is not always clear whether the energy required to produce the replacement heifer should be charged to milk production. If not, this requirement may approximately balance the output of beef from cull cows and surplus bull calves.

Sheep have similar requirements for production and compete for the same feed resources as beef cattle. Energy use for lamb production in Colorado is similar to beef production as estimated by Gee and about the same as pork and broilers on a liveweight basis. However, sheep produce both meat and wool and energy inputs should be apportioned to these products. A possible result of increasing energy costs is a change from synthetics to wool fabrics, because the energy required from petroleum products to produce synthetic fabrics is estimated to be two to three times that for wool, (Cook, et al., 1976).

Horses are major consumers of forages in this country and consume more forage than sheep. Horses, however, are almost entirely used for pleasure now and energy inputs have not been evaluated in terms of a tangible product.

Energy use in beef production can be divided into two general categories: fuel used directly for the operation of livestock ranches including transportation of cattle and feed, and the energy used in the feedlot, primarily energy embodied in feed. The major energy use by cow-calf operations on the extensive ranges of the West are for pickup trucks and producing hay for winter feed. Table 13.4 illustrates the percentage of total energy use by the Colorado ranching operations (Cook, et al., 1980). Fuel use by motor vehicles per pound of gain taken from Colorado ranch management records indicates that 7,200 to 13,000 BTU of fuel was required per pound of gain from cow-calf operations. Range production of beef is no longer a low energy operation since the horse has been replaced by the pickup truck and the cattle drive by transport trucks. Cow-calf enterprises in the Southeastern U.S., by contrast, require less land per cow and require less transportation; but pasture is generally more dependent on nitrogen fertilizer, which represents the major energy input in that region.

Several studies of energy requirements for cow-calf operations on Colorado ranches have been presented by Ward (1980); Ward, et al.; and Cook, et al., (1976). These values average about 9,000 BTU/lb of liveweight gain for

Table 13.4
Distribution of energy consumption in average beef production

Item	Within Items	Among Items
	-------- Percent	--------
Cow-Calf Production		21.2
Pickup and Trucks	41.3	
Machinery	23.9	
Fence	15.4	
Feed	10.0	
Water Facilities	6.2	
Labor	3.5	
	100.0	
Stocker (400-700 lbs)		16.0
Feedlot (700-1,100 lbs)		62.8
Feed Production	86.4	
Feed Processing	3.5	
Feedlot Operation	10.1	
	100.0	
		100.0

calves (Table 13.5). Estimates for California (Hertlein; Cervinka, et al.) are two to three times higher than the Colorado results, and one study of New Mexico ranches is considerably higher than for California (Patrick). Walker, et al. found values of 7,800 to 55,000 BTU/lb of gain ranging from unimproved to heavy fertilization. Since calves are seldom sold for slaughter, the energy inputs cannot be related to meat except in the case of veal calves.

Yearling cattle (between one and two years of age) are commonly supported on range forage or pasture for a few months to a year after which some are slaughtered and a majority go to feedlots for finishing on high-grain diets. Estimates of the energy use by yearling cattle in Colorado are 9,400 BTU/lb for mountain summer pastures, 10,000 for the High Plains, and 24,000 for irrigated pasture (Ward, et al.).

Table 13.5
Energy used in beef production

Item	Energy per Unit Product (10^3)BTU/lb	Energy per Unit Protein (10^3)BTU/lb
COW-CALF LIVE WEIGHT GAINS:		
Colorado Ranch[a]	9.84	
Colorado Ranch[b]	17.58-10.13	
New Mexico Ranch[c]	57.08	
California Ranch[d]	21.68	
California Ranch[e]	38.48	
Alabama Pasture[i]	3.13	
YEARLINGS LIVE WEIGHT GAINS:		
Alabama[i]	14.92-16.62	91.94-93.39
Colorado Ranch[a]	9.34	
Colorado Ranch[b]	9.81	
Colorado Irrigated Pastures[a]	24.21	
Midwest Pasture[f]	.43	
TOTAL BEEF PRODUCTION CARCASS BEEF:		
U.S. Average[g]	7.23-46.97	
Colorado[a]	20.23-46.06	26.19
U.S. Average Feedlot[f]	--	143.06
U.S. Average Range Fed[f]	--	18.24
U.S. Average[h]	39.20	

[a] Ward, 1980.
[b] Cook et al., 1980.
[c] Patrick.
[d] Hertlein.
[e] Cervinka et al.
[f] Pimentel.
[g] Lockeretz.
[h] Pierotti.
[i] Hoveland.

Feedlot finishing is more energy intensive than the other phases of beef production as indicated in Table 13.4. Feed is estimated to be 63 percent of total energy use for beef cattle production. Energy calculations for this phase are easily divided into two segments, feedlot operations and feed. Feedlot operation includes primarily feed processing and delivery of feed, but also corral cleaning and manure removal. An energy credit for the fertilizer value of manure could be claimed but has not been included in the estimates published. The feedlot phase represents 63 percent of the total energy use, of which more than 80 percent is for feed production. Despite heavy mechanization, feed processing and feedlot operations constitute a small percentage of the total. The fact that so much of the energy input is in the form of feed means that growth stimulators that increase feed efficiency are making an important contribution to energy conservation.

Since 63 percent of the energy is used in the feedlot phase of beef production, it might seem that energy conservation could be achieved by producting beef on forage only. There are two problems with this approach. First, forage finishing of beef requires more time and a higher feed intake for maintenance; and second, because such cattle are older when they reach the same weight and beef quality characteristics, they are less desirable. Colorado State University data (Table 13.6) indicate that feedlot-finished beef required nearly twice as much energy per animal as did grass-fed, but only 50 percent more per pound of beef produced. More cows must be supported to produce the same amount of beef under a system of forage only.

ENERGY MODELING OF LIVESTOCK PRODUCTION

Since the oil embargo of 1973, there has been increasing concern about the impact of rising energy costs on food prices and the degree of efficiency in agricultural production practices. National, state and regional, and firm level studies have estimated potential impacts from rising energy prices and constraints on the availability of fossil fuels.

Carter and Youde emphasize that the central energy question facing U.S. agriculture is more price related than supply related. They describe two price effects on agriculture. First, there are direct price increases pertaining to increases in energy based farm inputs (e.g., fuel,

Table 13.6
Energy efficiency of grass vs. feedlot beef production

Item	Range Fed	Feedlot 97 Days
Fossil Energy Input (BTU)	8.53×10^6	1.76×10^7
Slaughter Weight (lb)	862	1,166
Carcass Weight (lb)	462	723
Protein (lb)	157.7	192.3
BTU/lb Carcass	1.84×10^4	2.44×10^4
BTU/lb Protein	5.40×10^4	9.18×10^4

Source: Ward, 1980.

electricity, fertilizers, and chemicals). Second, there are indirect price increases resulting from the impact of rising energy prices on the general price level.

Another distinction should be made regarding types of energy. The measurement of total energy accounts for both direct energy use and indirect energy use (Hannon, et al.). Direct energy is associated with fuels for propulsion and utilities. Indirect or embodied energy includes energy embodied in inputs such as tractors, barns, and manufactured fertilizers and chemicals.

Agricultural production at the farm level accounts for only 3 to 4 percent of the total U.S. consumption of direct energy. However, the production, processing and distribution of agricultural products accounts for nearly 20 percent (Havlicek and Capps). Adding embodied energy to direct energy used at the farm, wholesale and retail levels gives total energy consumption.

According to the 1974 and 1978 energy data bases, energy used in livestock operations accounts for 11 percent of the energy used in crop and livestock production (see Table 13.7). Between 1974 and 1978, energy used in livestock operations increased 12.2 percent. On a per head basis, the major energy users are cow-calf, dairy, and feedlot operations. In terms of energy use per dollar of livestock production, turkey and broilers are the most intensive users of energy.

Table 13.7
Energy use for livestock and crop production in the U.S., 1974 and 1978[a]

Item	1974	1978	% Change 1974-78
	---------- Billion BTU	----------	
Livestock Operations	207,810	233,179	12.2%
Crop Operations	1,711,428	1,824,843	6.6%
TOTAL	1,919,238	2,058,022	7.2%

Source: Energy and U.S. Agriculture: 1974 and 1978. U.S. Department of Agriculture, ESCS Statistical Bulletin No. 632, 1980.

[a]Data include all energy used directly on the farm for crop and livestock production.

There are several statistical and mathematical modeling techniques used to estimate impacts of energy price increases and availability. Most often used are econometric and simulation models, input-output models, and mathematical programming models. In the following sections, we review two previous studies on energy modeling of livestock production accomplished as part of the Western Regional Project W-140.

Results from Two Livestock Studies

Out of this region research effort come two studies measuring energy impacts on the U.S. livestock industry. The first study examines the effects of energy price increases on all major livestock products. The other study evaluates energy conservation measures on U.S. beef production.

Yanagida and Conway used the livestock model of the U.S. Department of Agriculture to assess the effects of increasing energy prices on production and prices of beef, pork, chicken, turkey, eggs, and dairy. The model is recursively and simultaneously solved for an equilibrium solution by the Gauss-Seidel iterative technique (see Heien).

Two energy price increases were chosen for the study--12.5 percent and 25 percent. The former is consistent with average yearly price increases in the 1970s. The latter is a high price impact. These energy price impacts were assumed to occur continuously for four years.

The impact of energy prices is incorporated in three sections of the livestock model. First, higher energy prices result in higher feed grain and soybean meal prices. Second, higher energy prices affect the production process at the farm and ranch level (e.g., feed processing and distribution, waste handling, farm machinery, ventilation, space heating, etc.). Third, higher energy prices also affect the processing and transportation of livestock.

The model results indicate that energy price increases will have substantial and differential impacts on livestock prices and production (see Table 13.8). The poultry and egg sectors have the highest price and quantity impacts. For instance in the 25 percent scenario, the model results show poultry and egg prices in the fourth year (1982) to be between 14 and 19 percent higher than the USDA baseline while beef, pork, and dairy prices are expected to be between 7 and 8 percent above the baseline. Using a multi-period framework enables the livestock model to capture lagged production responses. Higher prices and lower levels of production did occur for all livestock commodities in the third and fourth years.

Yorks, et al., developed a U.S. beef production model to evaluate energy conservation measures induced by increased energy costs. The first or baseline scenario, designated as A, was required to produce the same quantity and quality of beef as produced in 1976 at minimum cost. This amounted to 12 million tons, 43 percent USDA grades choice and prime, 42 percent grades good or standard, and 15 percent ungraded. Scenario B was required to produce the same quantity of beef but with no constraint upon quality or grade. Scenario C was the same as A except that energy use was minimized instead of cost. Scenario D was the same as B except that energy instead of cost was minimized.

Feeding choices selected by each scenario of the Yorks, et al. study are listed in Table 13.9. The North Plains, South Plains and West are areas where irrigated

Table 13.8
Effects on livestock prices and production for 12.5 percent and 25 percent increases in energy prices (percentage changes from baseline)

Retail Prices	1979		1980		1981		1982	
	12.5%	25%	12.5%	25%	12.5%	25%	12.5%	25%
Prices								
PORIR.67	0.21	1.05	1.38	2.93	2.77	5.54	5.47	7.81
BEEIR	-0.22	1.07	-0.04	1.25	0.75	3.88	3.31	7.55
CHIIRFR	1.00	3.47	3.65	6.49	6.67	9.42	13.90	14.35
TURPR	0.81	3.89	3.76	6.62	7.58	10.75	15.97	17.09
EGGIR.67	0.54	3.96	2.59	7.70	5.68	13.84	11.49	18.64
DAIIR	1.00	1.77	1.87	3.01	2.71	5.36	4.64	8.23
Production								
PORAP-77	-0.13	-0.65	-0.70	-2.71	-1.31	-3.48	-1.88	-3.53
BEEAP	-0.13	-0.80	-0.25	-1.99	-0.48	-2.04	-1.95	-4.14
CHISPYO	-0.42	-0.74	-1.19	-2.68	-1.85	-3.49	-3.41	-6.11
TURAP	-0.77	-1.67	-2.95	-4.68	-4.53	-8.27	-7.72	-9.68
EGGAP	-0.47	-2.27	-1.84	-5.26	-3.84	-7.07	-7.55	-8.00
MILAP	-0.48	-0.76	-0.67	-1.41	-1.00	-1.99	-1.59	-2.96

Source: Yanagida and Conway.

Variable Names and Descriptions

PORIR.67	Retail pork price index	1967=1.0
BEEIR	Retail beef price index	1967=1.0
CHIIRFR	Retail frying chicken price index	1967=1.0
TURPR	Retail turkey price	cents/lb
EGGIR.67	Retail egg price index	1967=1.0
DAIIR	Retail dairy products weighted price index	1967=1.0
PORAP-77	Pork production, carcass weight	Mil. lbs.
BEEAP	Total beef production, carcass weight	Mil. lbs.
CHISPYO	Young chicken production, ready-to-cook	Mil. lbs.
TURAP	Turkey production, ready-to-cook	Mil. lbs.
EGGAP	Egg production	Mil. doz.
MILAP	Total milk production	Bil. lbs.

Table 13.9
Comparison of beef feeding activities

REGION Feeding Activity		Model A	Model B	Model C	Model D
		-- Animal Numbers x 10^6 --			
CORN BELT					
Heifers:	Corn Silage Corn	-	-	8.23	3.26*
Heifers:	Corn Silage Only	-	-	-	0.74*
Steers:	Supp. Grazing Corn	-	-	8.97	-
Heifers:	Supp. Grazing Corn	5.50	5.50	-	-
SOUTHEAST					
Heifers:	Corn Silage Corn	-	-	3.75*	-
Heifers:	Corn Silage Only	-	-	0.09*	-
Steers:	Supp. Grazing Corn	-	-	4.83*	-
Heifers:	Supp. Grazing Corn	-	-	-	3.11
NORTH PLAINS					
Heifers:	Supp. Grazing Only	3.01*	3.32	-	-
Steers:	Supp. Grazing Corn	8.34*	6.04	-	-
Heifers:	Supp. Grazing Corn	-	-	-	-
SOUTH PLAINS					
Steers:	Supp. Grazing only	-	-	-	4.40*
Heifers:	Supp. Grazing Only	-	-	-	2.93*
Steers:	Supp. Grazing Corn	3.41*	2.81*	-	-
Heifers:	Supp. Grazing Corn	2.93*	2.93*	-	-
WEST					
Heifers:	Corn Silage	-	-	-	0.54*
Steers:	Supp. Grazing Only	0.32	-	-	4.99*
Heifers:	Supp. Grazing Only	-	-	-	3.58*
Steers:	Supp. Grazing Corn	0.04*	-	-	-
Heifers:	Supp. Grazing Corn	1.10*	0.79*	-	-

Source: Yorks, et al.

[a]Denotes activity limited by supply of one or more feedstuffs.

feed crop production is significant. Hence, with energy minimization scenarios, feedlot finishing is eliminated in these areas, with the exception of corn silage use. Cost minimization causes feedlot finishing to be distributed regionally in a pattern similar to current practice.

The tradeoffs between cost and energy for restrained quality grade production or for unrestrained production indicate that energy conservation will be more costly. The cost differential to maintain current quality grade distribution was also shown to increase as energy input to the system is reduced. The unrestrained scenario minimizing cost, produces a quality grade breakdown similar to that in the present actual situation. This can be interpreted to mean that present grade relationships are justified on the basis of feed costs, even without consideration of the premium received for the higher grades.

Cost and energy efficiency show an inverse relationship. In Scenario A, the least cost approach, cost for feed is indicated to be $0.57 per pound of carcass beef produced. In Scenario D, energy utilization is reduced by 11 percent but cost increases by 6 percent. A change from Scenario A to D, on a national scale, represents an energy reduction of 4 by 10^{12} BTU for the production of 20.5 billion pounds of carcass beef.

TECHNOLOGICAL RESPONSE FOR THE FUTURE

The future for livestock production in the western states is heavily dependent upon irrigation and its supply of feed. The cattle feeding and dairy operations cannot be supported without the feed supplies provided by irrigated lands. Reductions of irrigated forage production could significantly reduce the complementary output of beef and sheep from western grazing lands. Likewise, since animal production has been the primary user of feed from irrigated land in most of the West, any decline in livestock production would have a marked impact upon irrigated agriculture.

Demand for animal products in the future, of course, is a factor in determining any technological changes. Beef demand, the major animal product, has been declining in recent years. This is a combination of changes in the cattle cycle, a prolonged recession and an unquantified trend toward non-meat diets. The options that are available to livestock producers, particularly beef producers, in response to energy and water shortages have been detailed in a publication by Ward (1980). A severe limitation on the quantity of irrigation water available would certainly have a severe impact on beef production in the arid West. The fact seems inescapable that under this situation the humid farming regions, particularly those of the Midwest, would have an economic advantage that would curtail cattle feeding and possibly severely limit milk production for the fluid market. The result might be a return to use of the grazing lands only to support breeding herds and for the slaughter of these animals off grass or with their shipment to the Midwest for finishing.

REFERENCES

Carter, Harold O. and James G. Youde. "Some Impacts of the Changing Energy Situation on U.S. Agriculture." Amer. J. Agr. Econ., 56: 878-87, 1974.
Cervinka, V., W.J. Chancellor, R.J. Coffelt, Curley, and J.B. Dobie. Energy Requirements for Agriculture in California. Joint Study, California Department of Food and Agriculture and University of California, Davis, 1974.
Cook, C.W., A.H. Denham, E.T. Bartlett, and R.D. Child. "Efficiency of Converting Nutrients and Cultural Energy in Various Feeding and Grazing Systems." Journal Range Management, 29: 186, 1976.
Cook, C.W., J.J. Combs, and G.M. Ward. "Cultural Energy in U.S. Beef Production." In Hand of Energy Utilization in Agriculture, Ed. David Pimentel, CRC Press, Boca Ratan, Florida, pp. 405-48, 1980
Coppock, C.E., D.L. Bath, and B. Harris. "From Feeding to Feeding Systems." Journal Diary Science, 64: 1230, 1981.
Dvoskin, D., E.O. Heady, and B.C. English. Energy Use in U.S. Agriculture: An Evaluation of National and Regional Impacts from Alternative Energy Policies. Center for Agricultural and Rural Development, Iowa State University, CARD -78, 1978.
FEA/USDA. Energy and U.S. Agriculture: 1974 Data Base, Volume I, FEA/D-76/459, FEA Office of Conservation & Environment, Washington, D.C. 20461, September 1976.
Gee, C.K. "A new Look at Sheep for Colorado Ranchers and

Farmers." Colorado State Univ. Experiment Station General Series 981, 1979.

Hannon, B., R. Herendeen, R. Puelo, and A. Sebald. "Energy Employment and Dollar Impacts on Alternative Transportation Option." In The Energy Conservation Papers, Ed. Robert H. William, Cambridge: Ballinger Publishing Co., pp. 105-29, 1975.

Havelicek, Joseph Jr. and Oral Capps Jr. "Needed Research with Respect to Energy Use in Agricultural Production." Southern Journal of Agricultural Economics, 9: 1-8, 1977.

Heien, D., J. Mathews, and A. Womack. "A Methods Note on the Gauss-Seidel Algorithm for Solving Economic Models." Agricultural Economics Research. 23: 71-80, 1978.

Hertlein, J.R. "Energy use by California Livestock." Report from Department of Animal Sciences, University of California-Davis, 1974.

Hoveland, C.S. "Energy Inputs for Beef Cattle on Pasture." In Handbook of Energy Utilization in Agriculture, Ed. David Pimentel, 1980.

Kliebenstien, James and Jean-Paul Chavas. "Adjustments of Midwest Grain Farm Businesses in Response to Increases in Petroleum Energy Prices." Southern Journal of Agricultural Economics, 9: 143-148, 1977.

Lasley, F.A. "Fuel and the Cost of Food." Journal Northeastern Agricultural Economics Council, 3: 46, 1974.

Lockeretz, William. Agricultural Resources Consumed in Beef Production. Center for the Biology of Natural Systems, Washington University (St. Louis), 1975.

Maki, Wilbur R. and Peter C. Knobloch. Regional Impacts of Alternative Energy Allocation Strategies. Department of Agricultural and Applied Economics, University of Minnesota, p. 74-8, 1974.

Miranowski, John. "Effects of Energy Price Increases, Energy Constraints, and Energy Minimization on Crop and Livestock Production Activities." North Central Journal of Agricultural Economics, 1: 1-14, 1979.

Patrick, Neil A. "Energy Use Patterns for Agricultural Production in New Mexico." Paper presented at the Energy and Agricultural Conference, St. Louis, Missouri, 1976.

Penn, J.B., B.A. McCarl, L. Brink, and G.D. Irwin. "Modeling and Simulation of the U.S. Economy with Alternative Energy Availabilities." American Journal of Agricultural Economics, 58: 663-71, 1976.

Pierotti, A., A.G. Keeler, and A.J. Fritsch. Energy and Food. CSPI Energy Series X, Center for Science in the Public Interest, Washington, D.C., 1977

Pimentel, D. Ed. Handbook of Energy Utilization in Agriculture. CRC Press, Boca Ratan, Florida, 1980.

Pimentel, D., W. Dritschilo, J. Krummel, and J. Kutzman. "Energy and Land Constraints in Food Protein Production." Science, 190: 754-761, 1975.

Roberts, Roland K., Gary R. Vieth, and James C. Nolan, Jr. "An Analysis of the Impact of Energy Price Escalations During the 1970's on Hawaii Beef Production and Prices." Contributed Paper, American Agricultural Economics Association Annual Meetings, Logan, Utah, 1982.

Stewart, B.A. "Irrigated Grain Production in the High Plains as a User and Producer of Energy." Energy in Agriculture-Use and Production. Crop Production and Utilization Symposium, Amarillo, February 14, 1980.

U.S. Department of Agriculture, ESCS. Energy and U.S. Agriculture: 1974 and 1978. Statistical Bulletin No. 632, 115 pp., 1980.

Walker, J.N., O.J. Lower, G. Benock, N. Gay, and E.M. Smith. "Production of Beef with Minimum Grain and Fossil Energy Inputs." Final Report National Science Foundation Grant AER-75-18706, Univ. of Kentucky, 1977.

Ward, G.M. "Energy, Land and Feed Constraints in Beef Production in the 80's." Journal Animal Science, 51: 1051-1064, 1980.

Ward, G.M. "Economic and Resource Costs of Production by Foods in Ruminants." In Animal Products in Human Nutrition, Iowa State University, Academic Press, New York, 1982.

Ward, G.M., P.L. Knox, B. Hobson, and T.P. Yorks. "Energy Requirements of Alternative Beef Production Systems in Colorado." In Agriculture and Energy, William Lockeretz (Ed.), Academic Press, New York, p. 395-411, 1977.

Whittlesey, Norman K. "Irrigation Development in the Pacific Northwest: A Mixed Blessing." Environmental Law, Northwestern School of Law of Lewis & Clark College, Portland, Winter 10(2): 1980.

Yanagida, John F. and Roger K. Conway. "The Effect of Energy Price Increases on the U.S. Livestock Sector." Canadian Journal of Agricultural Economics, 29: 295-302, 1981.

Yorks, T.P., W.C. Miller, J.J. Combs, and G.M. Ward. "Energy Minimized or Cost Minimized Alternatives for the U.S. Beef Production System." Agricultural Systems, 6: 121, 1980.

14
Optimization of Irrigation System Design

Edwin B. Roberts
Robert M. Hagan

QUANTIFICATION OF COSTS

The adoption of new or modified irrigation systems by farmers is generally determined by the economic efficiency of these systems rather than the energy use efficiency. Therefore, as we are seeking alternative irrigation systems with reduced energy requirements, we must keep in mind the economic costs. This section briefly discusses some procedures for calculating irrigation system costs.

METHODS OF ANALYSIS

Life-Cycle Costs

Probably the best way to compare costs of alternative irrigation systems is by calculating life-cycle costs for each. Life-cycle costing sums the operating cost of a system over its life cycle together with the net costs of purchase and installation (less any salvage value), maintenance, repair, and replacement. It includes the cost of money over the life of the system. The preferred alternative is the system with the lowest total life-cycle cost.

The life cycle is the period of time between the starting point and the cutoff date of analysis, over which the costs and benefits of a certain alternative are incurred. It can be the same as the useful life of the equipment or it can be chosen arbitrarily.

A general formula for life cycle costs of irrigation systems is as follows:

$$(14.1)\quad LCC = P - S + M + R + L + E$$

where:

LCC = life-cycle cost
P = purchase and installation costs
S = salvage value
M = maintenance and repair cost
R = replacement cost
L = labor cost
E = energy cost

Life-cycle costs can be expressed as either present value or annual value dollars. Present value is defined as the equivalent value of past, present, and future dollars corresponding to today's values. Annual values are past, present, and future costs converted to an equivalent constant amount recurring annually over the evaluation period. Past and future costs are converted to present values using the process of discounting.

Discounting and Calculation of Present Value

Discounting is the process by which costs occurring at different times during the life-cycle are put on a time equivalent basis to take into account the time value of money. The time value of money reflects the fact that money can be invested over time to yield a return over and above inflation. If $100 is invested now at an annual interest rate of 10 percent, the value one year from now will be $110. Thus, a cost of $110 one year from now has a present value of $100. The present value of a future cost is calculated using the single present worth formula:

$$(14.2)\quad P = F\left(\frac{1}{(1 + i)^n}\right)$$

where:

P = present value (or present worth)
F = future value
i = discount rate (or interest rate) as a decimal
n = number of years

The present value of \$110 one year from now using a discount rate of 10 percent is:

$$P = 110\left(\frac{1}{(1 + .10)^{1}}\right) = \$100$$

The present value of a uniform series of costs is calculated using the uniform present worth formula:

$$(14.3)\quad P = A\left[\frac{(1 + i)^{n} - 1}{i(1 + i)^{n}}\right]$$

where:

P = present value
A = an end-of-period payment in a uniform series of payments over n periods
i = discount rate
n = number of years

For example, assume that a cost of \$100 is paid out at the end of each year of a 10-year period of evaluation. Calculate the total present value of the 10 payments using a discount rate of 10 percent.

$$P = 100\left[\frac{(1 + .10)^{10} - 1}{.10(1 + .10)^{10}}\right] = \$614.46$$

If the annual costs are escalated, a modified uniform present worth formula is used:

$$(14.4)\quad P = A\left(\frac{1 + e}{i - e}\right)\left(1 - \frac{1 + e}{1 + i}^{n}\right)$$

where:

P = present value
A = an end-of-period payment in a uniform series of payments over n periods
i = discount rate
n = number of years
e = rate of escalation of A in each of n periods

In the previous example, the annual cost paid at the end of each year was \$100 and the present value of the 10 annual payments was \$614.46. If an annual escalation rate of 5 percent is used, the annual payment at the end of the first year is:

$$\$100 \times (1 + .05)^1 = \$105$$

At the end of the second year, the annual payment is:

$$\$100 \times (1 + .05)^2 = \$110.25$$

The present value of the 10 escalating payments is:

$$P = 100 \left(\frac{1 + .05}{.10 - .05}\right)\left(1 - \frac{1 + .05}{1 + .10}^{10}\right) = \$781.18$$

Effects of Inflation

The effect of inflation on life-cycle cost calculations is demonstrated with some very simple examples. Assume that an irrigation system has an initial cost of \$50,000, a useful life of five years, and the annual operating cost is \$10,000 in current dollars. Each component in the operating cost (maintenance and repair, replacement, labor, and energy) is escalating at the same rate, equal to the approximate general rate of inflation, assumed to be 5 percent. Assume that the discount rate and the approximate cost of borrowing money are 15 percent. Calculate the present value of the irrigation system using a three-year period of evaluation (life cycle).

The first step is to prepare a cash flow table, as shown in Table 14.1. The initial cost is shown in year zero, the start of the analysis. Because the system has a useful life of five years, a salvage value remains at the end of three years. Assuming a straight line depreciation, because two-fifths of the useful life remains, the salvage value is two-fifths of the initial cost, or \$20,000 in current dollars. However, because of inflation the current dollar value at the end of two years will be \$23,153. The present value of the salvage value, at a 15 percent discount rate, is \$15,223. The annual operating cost increases each year because of inflation. Present values of each cost are calculated using the single present worth formula. The total present value is \$59,855.

Table 14.1
Cash flow table for example irrigation system in current dollars showing effects of inflation

Item	Year			
	0	1	2	3
	---------- dollars ----------			
(Initial Cost: I)	(50,000)			
Present Value	50,000			
(Salvage Value: S)				(-23,153)
Present Value				-15,223
(Annual Operating Cost)		(10,500)	(11,025)	(11,576)
Present Values		9,130	8,336	7,612
Summed Present Values[a]	50,000	9,130	8,336	-7,611
TOTAL PRESENT VALUE (LCC)				$59,855

[a]Figures in parentheses are not included in totals.

Assumptions
3-year period of evaluation
5 percent inflation rate
15 percent discount rate

When constant dollars are used, the discount rate must be inflation free. An inflation-free discount rate can be determined in the following manner: If the total discount rate is 15 percent, $1 now is worth the same as $1.15 in one year. If the inflation rate is 5 percent, inflation alone causes $1 now to be the same as $1.05 in one year. The net increase in value (not counting inflation) is determined by dividing 1.15 by 1.05, giving a result of 1.0952. Therefore, the inflation-free discount rate is 9.52 percent.

It makes no difference whether life-cycle costs are calculated in current dollars or constant dollars, as long as the appropriate discount rate is used in each case. It is usually most convenient to calculate irrigation system costs in constant dollars.

Effects of Taxes

Most irrigation systems are operated by individuals or corporations subject to taxes; and therefore, cost calculations must take into account the effects of income tax deductions for depreciation and operating expenses, as well as investment tax credits.

Farms subject to income taxes can deduct the value of depreciation. The dollar savings resulting from the deduction depends on the marginal income tax rate. For example, if the farmer's income tax bracket is 30 percent, every dollar deducted for depreciation saves an actual $.30. With straight-line depreciation the annual tax savings from depreciation is:

$$(14.5)\quad \text{Tax Savings} = \frac{(I - S)}{n} \times r$$

where:

I = purchase and installation cost
S = salvage value
n = useful life in years
r = marginal income tax rate (decimal)

Tax savings resulting from deductions for operating costs (maintenance and repair, labor, and energy) are calculated as follows:

$$(14.6)\quad \text{Tax Savings} = (M + L + E) \times r$$

The investment tax credit allows an income tax credit equal to 10 percent of the cost of investments with lives of seven or more years. If the life is five or six years the investment credit is 10 percent of two-thirds of the cost. With a life of three or four years the investment credit is 10 percent of one-third of the cost. Where the life is three years or less, no investment credit is allowed.

Table 14.2 demonstrates life-cycle cost calculations which take into account the tax deductions and tax credits.

The irrigation system is assumed to cost $50,000 and have a life of five years. Therefore the system is replaced at the beginning of the sixth year. The present value of the replacement is calculated using the single present worth formula. Because the system has a life of five years, the allowed investment credit is equal to 10 percent of two-thirds of the investment cost, or $3,333. The annual tax savings from depreciation is $4,000.

The total present value of the life-cycle cost is $95,211. If we had ignored the tax benefits and calculated the life-cycle cost using the before-tax discount rate of 9.52 percent, the total present value would have been $141,707. Clearly, taxes have a very large effect on cost calculations and net costs of irrigation systems will vary from farm to farm depending upon the tax situations of the farmers.

Some users may wish to convert total present value figures into equivalent annual costs. This is done with the uniform capital recovery formula:

$$(14.7) \quad A = P \frac{i(1 + i)^{m}}{(1 + i)^{n} - 1}$$

where:

A = an end-of-period payment in a uniform series of payments over n years
P = present value
i = discount rate
n = number of years

Using the example shown in Table 14.2, the annual cost of the system including tax effects is $12,760. If tax effects are ignored, the annual cost comes to $22,589.

For preliminary screening of alternative irrigation systems an approximation of the before tax annual cost of each system can be derived as follows:

1. Divide the initial cost of each irrigation system component by the useful life of that component to account for depreciation.
2. Multiply the interest (discount) rate by one-half the total investment cost to approximate the average annual cost of money over the life of the system.

Table 14.2
Calculation of life-cycle cost of an irrigation system including tax effects[a]

Cost Calculation	Initial Cost	Present Value
Initial Investment:	\$(50,000)	\$35,828
$P = I[\frac{1}{(1+i)^n}] = \$50,000[\frac{1}{(1+.05712)^6}]$		
Investment Credit for Initial Investment:	(-3,333)	-3,153
$P = -3,333[\frac{1}{(1+.05712)^1}]$		
Investment Credit for Replacement:	(-3,333)	-2,388
$P = -3,333[\frac{1}{(1+.05712)^6}]$		
Tax Savings from Depreciation:	(-4,000)	-29,846
Annual Tax Savings $= (\frac{I-S}{n}) \times r$		
$= (\frac{\$50,000 - 0}{5}) \times .40 = 4,000$		
$P = A \frac{(1-i)^n - 1}{i(1+i)^n}$ n		
$= -\$4,000[\frac{(1+.05712)^{10} - 1}{.05712(1+.05712)^{10}}] = -29,846$		
Annual Operating Cost (M + L + E):	(10,000)	74,616
$P = \$10,000[\frac{(1+.05712)^{10} - 1}{.05712(1+.05712)^{10}}] = 74,616$		

(Continued)

Annual Tax Savings from Operating Cost Deductions:	(-4,000)	-29,846

$10,000 x .40

$$P = -4{,}000\left[\frac{(1 + .05712)^{10} - 1}{.05712(1 + .05712)^{10}}\right] = 29{,}846$$

PRESENT VALUE OF LIFE CYCLE COST (LCC): $95,211

---

[a]Basic Information
Period of analysis = 10 years
Inflation rate = 0% (all costs in constant dollars)
Inflation free discount rate = i = 9.52%
Marginal income tax rate = r = 40%
After-tax inflation-free discount rate = i' = i(1-r) = 9.52(1 - .40) = 5.712%
Initial cost of irrigation system = I = $50,000
Life of irrigation system = n = 5 years
Salvage value at end of life = 0
Annual operating cost = M + L + E = $10,000

3. Add annual operating costs to the depreciation and interest rates.

For the example shown in Table 14.2, this procedure results in an average annual cost of $22,380 which is very close to the before tax cost of $22,589 developed previously. This simplified approach, with minor variations, has been used in most of the publications dealing with costs of irrigation systems. It is also the basis of most computer programs dealing with irrigation costs. Unfortunately, the simplified approach does not deal with tax effects and does not consider the escalation of energy costs (or other costs) faster than the general rate of inflation.

APPLICATION OF INTERACTIVE COMPUTER PROGRAMS

Computer programs for irrigation system design and cost evaluation have been developed at a number of universities. Many of these programs are designed to provide approximate irrigation cost figures as input for crop production cost budgets, and thus they do not easily lend themselves to comparison of a wide range of alternative systems to find the least cost alternative. It is difficult to design a single program that will handle all irrigation methods in satisfactory detail. Therefore, a series of separate microcomputer programs have been developed at the University of California at Davis for examination of border, level border, furrow, hand-move sprinkler, side-roll sprinkler, center-pivot sprinkler, and linear sprinkler systems. The following description of the program for side-roll sprinklers illustrates its application.

1. Before operation of the computer program begins, it is necessary to determine the general layout of the system. The dimensions of the field, the available water holding capacity of the soil, the maximum infiltration rate of the soil, the crop rooting depth, the annual evapotranspiration and peak evapotranspiration rate of the crop, the well location and static depth to water, and the expected irrigation efficiencies must be determined. For example, assume an 80-acre rectangular alfalfa field with dimensions of 1,320 feet and 2,640 feet. The soil is a sandy loam with the available water holding capacity of 1.5 inches/foot and a maximum infiltration rate of 1 inch/hour. The crop rooting depth is 10 feet, annual evapotranspiration is 48 inches and the peak evapotranspiration rate is 0.26 inches/day. The well is located in the middle of one of the long sides of the field. The static depth to water in the well is 150 feet. The expected irrigation efficiency is 80 percent.

2. The system is designed with a buried-plastic mainline extending along the side of the field both directions from the well. Side-roll laterals with a length of approximately one-quarter mile are connected to valves along the mainline. For the first trial the user must select a spacing of laterals along the mainline. Assume a 60 feet spacing. Dividing the 2,640 feet field length by the 60 feet spacing means that there are 44 positions for lateral lines along the mainline.

3. With an available water holding capacity of 1.5 inches/foot and a rooting depth of 10 inches, the total available water in the root zone is 15 inches. Assume that the system is designed to irrigate when 50 percent of the available water (7.5 inches) is used up.

4. Dividing 7.5 inch by the peak evapotranspiration rate (.26 inches/day) we find that the interval between irrigations at the peak of the irrigation season can be no longer than 28.8 days.

5. Dividing the 7.5 inches net application of water by the irrigation efficiency of 80 percent, we find that the gross application is 9.375 inches per irrigation. Dividing 9.375 inches by the maximum infiltration rate of 1 inch/hour, at least 9.375 hours are required to apply the necessary amount of water. Therefore, two sets/day can be irrigated and the set time can be anywhere between 9.375 and about 11.5 hours, allowing about a half hour to move the laterals.

6. With 44 positions along the mainline and two sets/day, one lateral is capable of irrigating the field in 22 days. Because this is less than the maximum interval of 28.8 days, one lateral is acceptable for the first run.

7. The mainline is divided into sections according to the maximum number of laterals served by each section. With one lateral, when the lateral is at the end position on the field, the mainline will have to carry the entire pump flow from the center of the field to the end. A subroutine of the computer program allows the user to input the number of mainline sections and the length of each. (It is usually helpful to prepare a series of rough sketches of the pipe layout for different positions of the laterals.)

8. When the program is run, the computer prompts the user for input as to the annual net water application (48 inches), the static depth to water (150 feet), typical drawdown in wells (21 feet/100 gpm assumed in this case), total acreage irrigated, maximum infiltration rate, peak evapotranspiration, net application (7.5 inches), irrigation efficiency, minimum set time, maximum set time, number of sets per day, spacing of laterals along the mainline, spacing of sprinklers along the laterals (usually 30 feet or 40 feet with side-roll sprinklers), the number of valves along the mainline (44), the number of laterals operating simultaneously (1), the number of sets per irrigation cycle (22), the length of the laterals, and the number of sprinklers per lateral (33 with a 1,320 feet field width and sprinkler spacing of 40 feet).

9. The computer then calculates the minimum and maximum application rates (inches/hour) corresponding to the minimum and maximum set times. The sprinkler discharges (gpm) that will provide the minimum and maximum

application rates to the area determined by the sprinkler spacing are then calculated.

10. The computer next prompts the user for the nominal diameter of the lateral lines (4 inches or 5 inches).

11. A data file currently includes 441 combinations of sprinkler pressure, discharge, and wetted diameter. These combinations are arranged in order of increasing pressure, from 8 to 80 pounds per square inch. The computer reads the first combination from the data file and checks to see if the discharge is between the limits set in step 8. It next checks to see if the wetted diameter is adequate for proper coverage with the chosen spacing of sprinklers along the laterals and laterals along the mainline. Whenever a combination does not satisfy the conditions, the computer skips to the next combination in the sprinkler file.

12. The computer multiplies the sprinkler discharge by the number of sprinklers per lateral to determine the total gpm of the lateral. for the nominal diameter selected in step 9, the computer selects the actual inside diameter and calculates the velocity and friction loss. If the velocity exceeds 5 feet/second, or the friction loss exceeds 15 percent of the design pressure, the computer skips to a new sprinkler.

13. The computer next selects a mainline pipe diameter, from a file incorporated into the program, for each section of the mainline so that the section can supply the maximum number of laterals served by that section with a water velocity of 5 feet/second or less. Then an adequate size is selected, the friction loss is calculated.

14. A calculation of the pumping head is made. This consists of the design pressure (converted to feet of head), friction loss in the lateral, friction loss in the mainline, the difference in elevation between the pump and the highest point of discharge, a miscellaneous head loss equal to 10 percent of the design pressure (to account for losses in elbows and couplings), the static depth to water, drawdown, and an allowance of 5 feet/100 feet for friction loss in the pump column.

15. The preliminary calculation of pump horsepower is made using the total head, the total pumping rate (gpm per lateral times the number of laterals), and an assumed bowl efficiency. The horsepower is used to select the required lineshaft diameter and calculate friction horsepower loss in the lineshaft. The total horsepower is determined by

adding the lineshaft horsepower loss to the preliminary horsepower.

16. The computer next selects an appropriate pump bowl model, from a file built into the program, and calculates the number of stages required. The sizes of the column, discharge head, electric motor, and starter panel are selected. The appropriate motor efficiency for the motor size selected is used to convert the horsepower into kilowatt demand.

17. The required number of irrigations per year is calculated and the annual usage in hours is determined by multiplying the set time by the number of sets per irrigation and the number of irrigations per year. Annual energy use is the product of the kilowatt demand and the annual hours.

18. Using the data built into the program, the computer calculates the total cost of all components, including the well cost and pump installation. Annual costs are calculated for depreciation, interest, property taxes and insurance, maintenance and replacement, labor, and energy (using any appropriate utility rate schedule).

19. The first calculation of overall annual cost is printed along with data about the design of the system. The annual cost and annual energy use are stored in test registers. The computer then selects another sprinkler model and repeats all calculations until another annual cost is obtained. The second annual cost is compared with the test register. If the second annual cost is more than 5 percent higher than the cost stored in the test register, the second system design is rejected and the computer chooses a new sprinkler. If the second value is higher than the test register but by no more than 5 percent the annual energy use is compared with that in the test register. If the energy use is higher than the test register the system is rejected. If it is lower than the test register, details of the second system design are printed. If the second design has a lower annual cost than the test register, the details of the second system are printed and new values are inserted into the test register.

20. After the computer has completed calculations with the file of the sprinkler models, the user may change the lateral pipe size, spacing along the lateral and mainline, the number of laterals, or the number of sets per day and repeat the process. While this procedure is tedious, it is capable of examining many thousands of combinations of sprinkler model, pressure and spacing, and

pipe size to select a very few alternative systems with near-minimum annual costs and pumping energy requirements.

REPLACEMENT OR MODIFICATION OF EXISTING SYSTEMS

When an irrigation system is being designed for a field that has not been previously irrigated, the designer has a relatively free choice of alternative systems (within constraints imposed by crop type, soil type, and topography). The costs of the alternative systems can be compared using any of the methods described above. However, when an existing irrigation system is being replaced or modified, the designer must take into account the investment in existing irrigation, land preparation, and harvesting equipment. This is done by calculating the present value for continued operation of the existing system for a certain number of years and comparing this with the present value for conversion to a new system and operation for the same number of years.

Components that have useful life remaining, but which will not be reused, may have a salvage value. However, some components, such as the bowls, column, and lineshaft for deep-well turbines may have no net salvage value because removal costs equal or exceed resale value. Costs not affected by the change in irrigation systems do not need to be included in the present value analyses. Tables 14.3 and 14.4 summarize the calculations involved in replacement of an existing side-roll sprinkler with a linear system.

Table 14.5 shows the effect of escalating energy costs on continued operation of the side-roll irrigation system. Energy is the dominant annual cost in all cases. Converting to the more energy efficient linear sprinkler system becomes cost effective at all levels of energy cost escalation from 2.5 percent upward. This is shown by comparing Tables 14.5 and 14.6.

Table 14.3
Calculation of present value for 10 years continued operation of existing side-roll sprinkler system (constant 1983 dollars)[a]

Description	Initial Cost	Present Value
Purchase and Installation Costs		$ 0
Total Salvage Value (end of 10th year)	$(-7,584)	-2,176
Annual Maintenance and Repair Cost	(1,676)	7,499
Replacement Costs:		
Sprinkler Heads (4th year)	(1,980)	
Pump Bowl Assembly (3rd & 10th years), each time	(3,707)	
Pump Installation (3rd & 10th years), each time	(1,760)	
Total After Taxes		8,156
Annual Labor Cost	(1,000)	4,477
Annual Savings from Depreciation Tax Deductions	(-1,545)	-11,528
Annual Energy Cost	(14,556)	65,169
TOTAL PRESENT VALUE		$71,957

[a]Basic information
Age of existing system = 5 years
Period of analysis = 10 years
Marginal income tax rate = 40%
After-tax inflation-free discount rate = 5.712%
Acres = 80, Crop = alfalfa, Field size = 1320' x 2640'
No. of sets per irrigation = 11
Annual no. of irrigations = 10
Set time = 23 hours, Annual hours of use = 2,530
Pump horsepower = 125, Demand = 86 kw, Annual kwh = 217,580

Table 14.4
Calculation of present value for conversion from side-roll sprinkler to linear sprinkler (constant 1983 dollars)[a]

Description	Initial Cost	Present Value
Purchase and Installation Costs:		
Linear Sprinkler		$ 53,240
New Pump Assembly, Motor, and Panel		13,695
Salvage Value (replaced components):		
Side-Roll Laterals		-15,840
Sprinkler Heads		- 743
Motor and Panel		- 4,017
Salvage Value (end of 10th year)	($21,767)	-12,490
Annual Maintenance and Repair Cost	(2,631)	11,777
Replacement Cost:		
Pump Bowls (7th year)	(2,916)	
Pump Installation (7th year)	(1,760)	
Total After Taxes		2,959
Annual Labor Cost	(400)	1,791
Annual Savings from Depreciation Tax Deduction	(4,984)	-14,878
Annual Energy Cost	(8,342)	$ 37,345
TOTAL PRESENT VALUE		$ 72,839

[a]Basic Information
Period of analysis = 10 years
Marginal income tax rate = 40%
After-tax inflation-free discount rate = 5.712%
Acres = 80
Crop = alfalfa
Field size = 1320' x 2640'
Annual hours of use = 2,671
Pump horsepower = 50
Demand = 44 kw
Annual kwh = 118,165

Table 14.5
Present values for continued operation of side-roll sprinkler systems with four different energy cost escalation rates

Description		Present Value
Ownership and Operating Cost (excluding energy)		$ 6,788
Energy Cost		
Escalation Rate Percent	= 0	65,169
	= 2½	73,997
	= 5	84,169
	= 10%	109,400
Total Present Value		
Energy Cost		
Escalation Rate Percent	= 0	71,957
	= 2½	80,785
	= 5	90,957
	= 10	$116,188

Table 14.6
Present values for conversion to linear sprinkler system with four different energy cost escalation rates

Description		Present Value
Ownership and Operating Cost (excluding energy)		$35,494
Energy Cost		
Escalation Rate Percent	= 0	37,345
	= 2½	42,404
	= 5	48,233
	= 10	62,692
Total Present Value		
Energy Cost		
Escalation Rate Percent	= 0	72,839
	= 2½	77,898
	= 5	83,727
	= 10	$98,186

15
Energy Intensity and Factor Substitution in Western United States Agriculture

Chennat Gopalakrishnan
John F. Yanagida

INTRODUCTION

Energy has been substituted for labor and other inputs for many years until U.S. agriculture is the most energy intensive in the world. Viewed against this background, the case for energy conservation and the increased physical efficiency of energy use becomes compelling. The underlying assumption here is that energy may be readily substituted by other inputs. However, any effort to replace energy by other factors of production (e.g., capital or labor) would entail redesigning processes and products and investing in energy-efficient equipment and technology. Changes in energy intensity provide a good index of such structural changes in production.

Empirical investigations using econometric models have been used to indicate the technical potential for substitution between energy and nonenergy inputs, especially in the manufacturing sector of the U.S. economy (Berndt and Wood). As more sophisticated techniques for estimating production and cost functions become available, the easier it will be to determine factor substitution possibilities.

This chapter examines energy intensity and factor substitution in western United States agriculture. The first section presents a conceptual framework of factor substitution. This is followed by a discussion of actual findings on changes in the energy intensity of western U.S. agriculture during 1974-1978. The final section presents evidence on the potential for energy-nonenergy input substitution in western U.S. agriculture.

FACTOR SUBSTITUTION: A CONCEPTUAL FRAMEWORK

The farm firm is an economic unit in which one or more commodities are produced. It is assumed that the farm manager buys inputs and sells commodities in competitive markets such that prices of inputs and outputs are determined by forces beyond the control of the manager and are regarded as given. In the following abstract analysis, production of one commodity is considered. Similar analyses can be extended to the production of two or more commodities.

The Production Function

The production function is defined as a relation between quantities of factors or inputs that a farm manager uses and the quantity of output produced. The farm manager transforms inputs into output or product, subject to the technical specifications of the production function. The production process typically requires a wide variety of inputs. These inputs are not as simple as "land", "labor" and "capital." Many qualitatively different types of inputs are used in production. Generally, two broad categories are used to classify inputs. Variable inputs are those factors whose level of use varies with the quantity of output produced in a give time period. Fixed inputs are those factors whose level of use does not vary with the level of output in a given time period. Examples of variable inputs are labor, fertilizer, gasoline, seed, etc. Land and buildings are commonly used examples of fixed factors of production.

Production of One Output with Two Variable Inputs

Assume that the production process for a commodity (Y), can be described as:

$$Y = f(X_1, X_2 | X_3, \ldots, X_n)$$

where:

Y = bushels of corn
X_1, X_2 = capital, energy
$X_3, \ldots, X_n$ = land, buildings, etc.

This production function states that different levels of corn can be produced with different combinations of capital and energy, given fixed levels of inputs such as land and buildings.

By holding output constant, we can illustrate the different combinations of capital and energy that can produce this level of output. The product curve or function depicting these input combinations is called an isoquant (see Figure 15.1). Here both input combinations A and B produce the same level of output Y_o. An isoquant map describes a set or family of isoquants each representing a specific level of output. These non-intersecting isoquants are potentially infinite in number, and reflect higher output levels with movements away from the origin. The slope of a line tangent to a point on the isoquant (which is equal to the slope of the isoquant at that point) shows the rate at which input X_1 may be substituted for X_2 (or *vice versa*) at the point of tangency, while maintaining production at that constant level.

There are three general cases of input substitution in production: perfect substitutes, perfect complements, and imperfect substitutes. Perfect substitutes are those inputs perfectly able to replace one another in production without affecting output. The shape of the isoquant, in this case, is a straight line from one axis to the other (Figure 15.2A). Resources that are perfect complements must be used only in fixed proportions. An additional amount of any one input will not change the level of output. In this case, the shape of the isoquant is a right-angle (Figure 15.2B). The third type of resource relationship, imperfect substitutes, is probably the most typical in agricultural production. For this case, successive incremental reductions in one input must be offset by increasingly larger amounts of the other input (Figure 15.2C).

The econometric approach to modeling production involves the development of production functions that show the quantity of output that can be obtained from given input quantities and how the level of output changes as the quantities of inputs change. Firms typically attempt to find the combination of factor inputs that minimizes the per unit cost of production. The profit-maximizing firm will take into account the prices of all inputs, while deciding the optimal input mix. When the price of one input rises in relation to the price of another, a firm will substitute the cheaper input to lower production

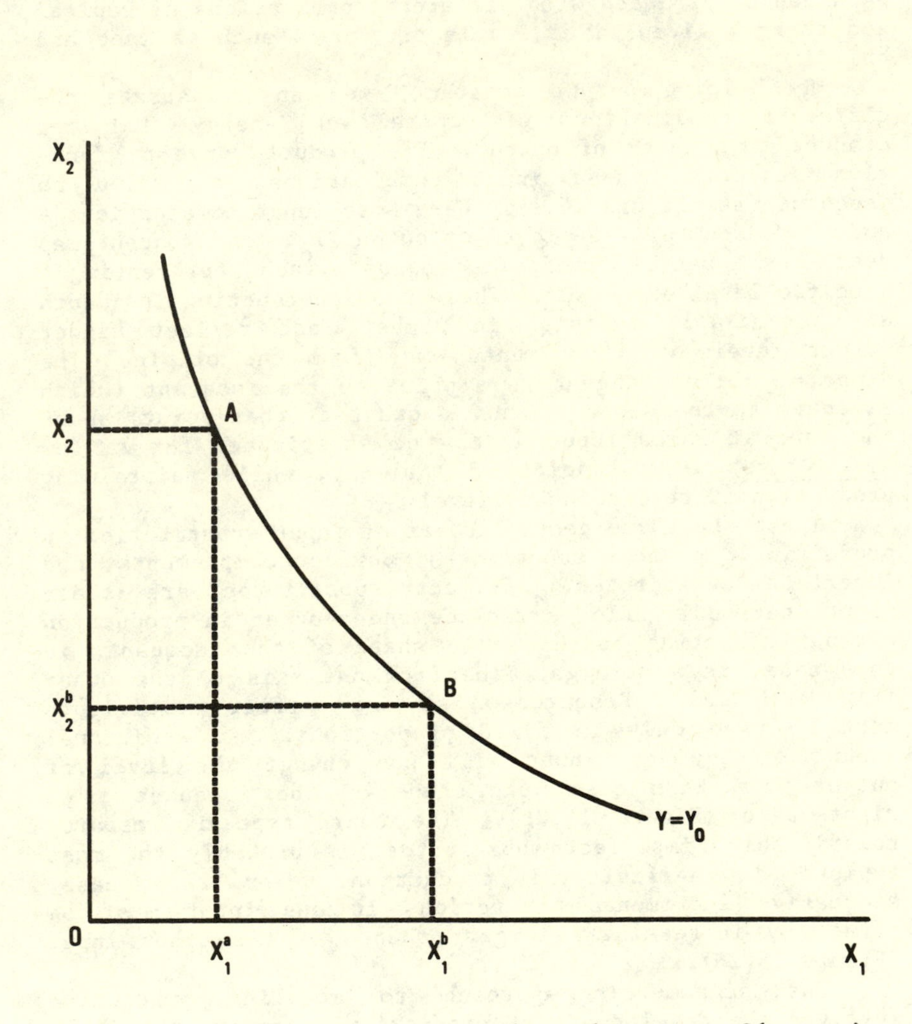

Figure 15.1 Production isoquant describing $Y_o = f(X_1, X_2)$

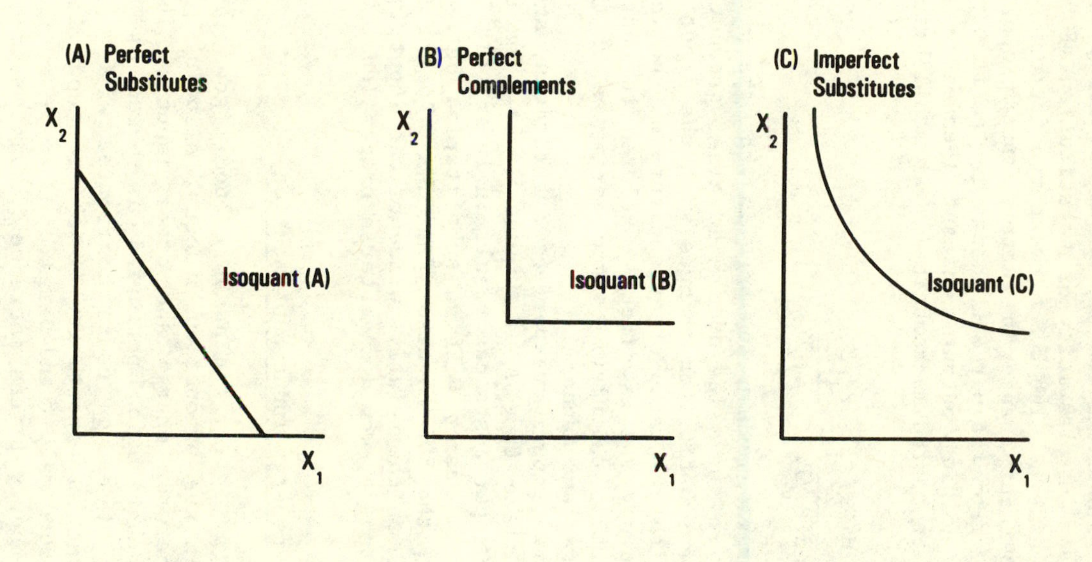

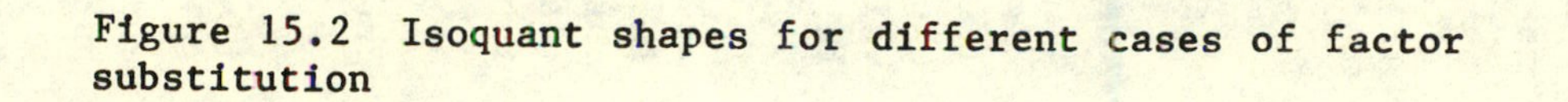
Figure 15.2 Isoquant shapes for different cases of factor substitution

costs. The rationale for factor substitution stems from this basic postulate in economics.

To measure the degree of substitutability among different inputs in the production process, economists use the concept of elasticity of substitution (EOS), generally symbolized as σ. The EOS between two factors of production is the percent change in factor input proportion (X_2/X_1) resulting from the percent change in the factor price ratio (P_1/P_2), holding output constant (Henderson and Quandt). This can be stated mathematically at equilibrium as:

$$\sigma = \frac{d(X_2/X_1)}{d(P_1/P_2)} \cdot \frac{P_1/P_2}{X_2/X_1}$$

The underlying assumption here is that the price of energy is equal to the value of its marginal product. The elasticity of substitution changes with the input's cost share (i.e., with the ratio of expenditure on the input to total expenditure), and is directly related to cost parameters, which enables calculation from estimates of these parameters (Hogan and Manne).

If two inputs are perfect substitutes, and the EOS is infinite (see Case A in Figure 15.3), small changes in relative input costs can result in infinitely large changes in input intensity. If no substitution between inputs is possible, the EOS is zero (Case B). If the EOS is unity (Case C), as the relative price of one input increases, its input intensity would register a proportionate decrease.

Energy-Nonenergy Input Substitution

Energy is treated as an economic good, as one of the inputs in the production process, allowing for a substitution between energy and other inputs. The extent of energy substitution potential is based on the observed behavior of energy users confronted with changing input prices and an opportunity to change input intensity patterns accordingly (Landsberg).

As energy prices increase, it is assumed that energy users will make use of all opportunities to employ energy-efficient machinery, and redesign products and processes to lower their output costs. Firms generally choose the combination of inputs which maximizes profits from available resources. Demand for each input will, therefore,

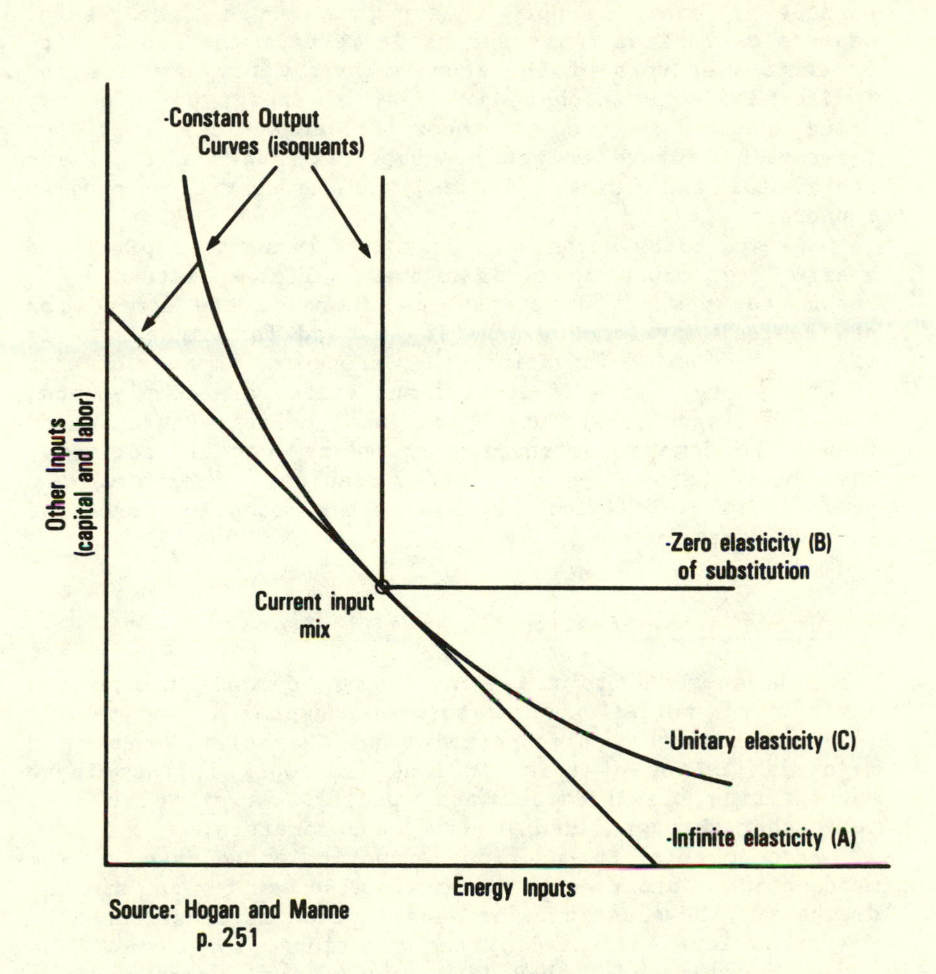

Figure 15.3 The elasticity of substitution concept

depend on the level or amount of output, the relative prices of all inputs, and the production technology.

The elasticity of substitution of energy as compared to other inputs determines the economic impact of large changes in energy supply and/or price. It is a useful measure or indicator insofar as it reveals the sensitivity of certain sectors of the economy to the complexities and indirect effects which a given set of energy policies may bring about. It is an important analytical tool for forecasting future energy consumption, and for assessing energy tax and price policies, and conservation program impacts.

As stated by Hogan, and Hogan and Manne if capital and energy are substitutes, national policies intended to reduce the demand for energy may increase the demand for new capital investment, creating a need for higher levels of saving, but mitigating the economic impact of the lowered energy use. If capital and energy are complements, and there is no substitution, national policies intended to reduce the demand for energy may reduce economic activity, thereby reducing the demand for capital, lessening the pressure on savings, but magnifying the economic impacts of lowered energy use.

Measuring Factor Substitution in Agriculture

One approach to measuring energy demand in American agriculture relies upon mathematical programming models (e.g., Miranowski; Kliebenstein and Chavas). However, a major limitation of this modeling framework is that input substitution possibilities are specified *a priori* by the researcher through fixed production parameters.

Econometric production function analyses, like mathematical programming models, also suffer a similar drawback. The elasticity of substitution is constrained to a certain level dictated by the functional form used. The commonly estimated Cobb-Douglas production function has a unitary elasticity of substitution between any pair of factors. The CES (Constant Elasticity of Substitution) production form has a constant elasticity of substitution for all pairwise input combinations. Unlike the Cobb-Douglas specification, the elasticity of substitution can be different from unity in the CES case.

Recently, flexible functional forms have been utilized allowing for the degree of factor substitution to be internally determined by the model. Studies by Berndt and Wood; Babin, et al.; and Miranowski and Mensah, have employed the translog function specification. This functional form offers the advantage of not imposing stringent conditions on the elasticity of factor substitution.

ENERGY INTENSITY IN WESTERN U.S. AGRICULTURE

This section presents and discusses information on energy intensity in western U.S. agriculture. The findings given here are based on information developed from data published by the Federal Energy Administration and the U.S. Department of Agriculture.

It is important to note that a generally accepted method for "measuring the energy intensity of an economy is to determine how much gross energy consumption is needed to produce a dollar's worth of output and then to trace this ratio through time" (La Bell). In this chapter, we have followed this approach in determining the energy intensity of agriculture.

Table 15.1 shows that energy used to produce a dollar's worth of agricultural output registered nearly a 9 percent decrease from 1974 to 1978, suggesting a reduction in the energy intensity of western U.S. agricultural production. A closer look at the table reveals that a decrease in energy intensity occurred in spite of an increase in the energy requirements for crop production, due to a significant reduction (almost 24 percent) in energy use in livestock production. The table also indicates that changes in energy intensity in western U.S. agriculture closely corresponded to energy intensity changes in U.S. agriculture as a whole.

What could have possibly triggered such a drastic reduction? In order to answer this question, it is useful to look at the pattern of energy use in U.S. livestock production. The major energy inputs in livestock production are feed handling (29 percent), feed processing and distribution (21 percent), farm travel (20 percent), and waste disposal (11 percent) (Stout). It is noted, however, that there could be wide variations in energy use, depending on the type of livestock operation. For instance, energy use for feed processing and distribution could be as high as 50 percent for a beef cow/calf operation and as low as 10 percent for a hog-farrow-to-finish operation. In

Table 15.1
Energy used to produce a dollar's worth of agricultural output, 1974 and 1978: U.S. and western U.S.[a]

Item	United States			Western United States		
	1974	1978	Percent Change	1974	1978	Percent Change
Agricultural Production (1,000 BTU)	24	22	-8.7	23	21	-8.7
Crop Production (1,000 BTU)	41	43	4.9	38	41	7.9
Livestock Production (BTU)	5,261	4,292	-18.4	4,882	3,724	-23.7

Source: Table developed from information given in *Energy and U.S. Agriculture: 1974 and 1978*, Statistical Bulletin No. 632 of ESCS, U.S. Department of Agriculture, 1980 and *Energy and U.S. Agriculture: 1974 Data Base* of FEA/US Department of Agriculture, Volumes 1 and 2, 1976 and 1977.

[a]Value of agricultural output is in constant 1974 dollars.

terms of energy consumed, gasoline tops the list (33.6 percent), followed by diesel fuel (30.6 percent), LP Gas (17.4 percent), electricity (15.2 percent), natural gas (2.3 percent), with fuel oil and coal together accounting for less than 2 percent.

It is possible to speculate about why energy intensity of livestock production has registered a sharp decline during the period under study. The wide-spread adoption, in recent years, of energy-efficient practices, especially in feed handling, feed processing and distribution, and waste disposal have contributed to significant energy savings. Also livestock producers have tended to switch to naturally ventilated, cold barns for beef housing and hog fattening, thereby eliminating energy used for supplemental heat and ventilation. A reduction in concentrates in livestock feed in recent years could also have contributed to energy savings in livestock production.

FACTOR SUBSTITUTION IN WESTERN U.S. AGRICULTURE

Data presented in the previous section clearly point to a decrease in energy intensity in western U.S. agriculture during the 1974-1978 period, when energy prices almost quadrupled. Intuitive logic suggests that the reduction in energy intensity could well be in part due to the substitution of lower priced inputs such as capital, land, and labor for the higher-priced units of energy. However, there are as yet no published studies which provide empirical evidence on the rate and extent of energy-nonenergy input substitution in western U.S. agriculture.

This section presents evidence on energy-nonenergy input substitution in western U.S. agriculture (Khaleghihoseinabadi). The findings are based exclusively on an analysis of secondary data. Table 15.2 gives estimates of own- and cross-price elasticities of demand for direct energy, indirect energy (fertilizer and other chemicals), capital, land, labor, and miscellaneous input. Estimates of Allen Partial Elasticities of Substitution (AES) for the same inputs are given in Table 15.3. AES measures the response of a 1 percent change in the relative price of an input to the percentage change in the ratio of inputs.

Own-price elasticity of demand, as estimated in Table 15.2 refers to the percentage change in quantity demanded

Table 15.2
Estimated own and cross-price elasticities of demand for energy, indirect energy, capital, land, labor, and other inputs

	Prices of					
Demand for:	Energy	Indirect Energy	Capital	Land	Labor	Miscellaneous Inputs
Energy	-0.8481	0.0749	0.4961	0.1119	0.1524	0.0128
Indirect Energy	0.0550	-0.3493	0.5518	0.6252	0.5020	0.0310
Capital	0.3213	0.4861	-0.9270	0.4540	0.6424	-0.0559
Land	0.0708	0.5380	0.4434	-1.1415	0.1182	-0.0288
Labor	0.1139	0.5103	0.7413	0.1397	-0.5811	0.1385
Miscellaneous Inputs	0.0242	0.0796	-0.1631	-0.0863	0.3501	-0.2045

Source: Khaleghioseinabadi, p. 97.

Table 15.3
Estimated Allen Partial Elasticities of Substitution for energy, indirect energy, capital, land, labor, and other input

Demand for:	Prices of					
	Energy	Indirect Energy	Capital	Land	Labor	Miscellaneous Inputs
Energy	-6.236	0.4046	2.3624	0.5205	0.8376	0.1779
Indirect Energy		-1.8879	2.6275	2.9079	2.0585	0.4302
Capital			-4.4143	2.1114	3.5298	-0.7764
Land				-5.3093	0.6496	-0.4012
Labor					-3.1930	1.9234
Miscellaneous Inputs						-2.8407

Source: Khaleghioseinabadi, p. 97.

of an input divided by the percentage change in its price, all else being equal. Likewise, cross-price elasticity (Table 15.2) refers to the percentage change in quantity demanded of one input divided by the percentage change in the price of another input, all else being equal. The findings suggest that the responsiveness of energy with respect to its own-price is about -0.85. A 10 percent increase in energy price will reduce energy consumption by 8.5 percent. The table also reveals that the elasticity of demand for indirect energy with respect to its own price is around -0.35.

The estimated cross-price elasticity of energy and capital suggests that a 10 percent increase in the price of energy would result in a 3.2 percent increase in capital use. The table also indicates that energy readily substitutes for labor. Thus, the cross-price elasticities for energy and labor show that a 10 percent increase in energy prices would raise labor use by 1.1 percent in farm production, suggesting a relatively insignificant energy-labor substitution. The cross-price elasticity of energy and land appears to suggest a low degree of substitution, since a doubling of energy price would result only in a modest increase of 7 percent in land use.

Table 15.2 further reveals that the degree of substitution between direct energy and indirect energy is quite insignificant. This makes sense since, in actual practice, the price of indirect energy is closely dependent on the price of fuel and any increase in fuel price usually leads to a rise in the price of its derivatives in the very near future, if not simultaneously.

Table 15.3 presents the estimates of Allen Partial Elasticities of Substitution (AES) for the different inputs used in western U.S. agricultural production. AES, as noted earlier, measures the percentage response of the ratio of input use to a one percent change in the ratio of input prices. Using Table 15.3, an elasticity of substitution of 2.36 for energy (E) and capital (K) implies that a one percent decrease in the price ratio (P_K/P_E) would result in a 2.36 percent decrease in the input ratio (E/K). Elasticities with a positive sign indicate substitution possibility between the two inputs and while negative elasticities denote complementarity between them.

Table 15.3 shows that the elasticity of substitution for each pair of energy and other inputs is positive, indicating that energy and the other factors of production are substitutes rather than complements. Among them, the substitutability between energy and capital is the highest, followed by energy and labor, energy and land, and energy and indirect energy.

The results presented in Tables 15.2 and 15.3 point essentially to the same conclusions regarding the degree and magnitude of energy-nonenergy input substitution in western U.S. agriculture: energy is substitutable with other factors of production; energy and capital have the highest potential for substitution, followed by energy and labor and energy and land; there is no complementarity between energy and the major nonenergy inputs.

CONCLUSIONS

The principal conclusion of this discussion is that the energy intensity of western U.S. agriculture registered a sharp decline during a period when oil prices in the United States nearly quadrupled. Evidently, the signals sent out by the market have had an impact on the pattern of resource use and the structure of agricultural production. While a closer look points to wide variations within sectors and regions, the conclusion is that western U.S. agriculture, like U.S. agriculture as a whole, has become less energy-intensive in the wake of sharp energy price escalations.

Another significant conclusion relates to energy-nonenergy input substitution in western U.S. agriculture. The findings presented point to the fact that energy is substitutable with other factors of production. Because of such substitution possibilities, there is a strong basis to conclude that U.S. agricultural production has the resilience to effectively cope with rising energy prices.

REFERENCES

Babin, Frederick G., Cleve E. Willis, and P. Geoffrey Allen. "Estimation of Substitution Possibilities Between Water and Other Production Inputs." Amer. J. of Agr. Econ., 64: 148-151, 1982.

Berndt, E.R. and D.O. Wood. "Technology, Prices and the Derived Demand for Energy." Review of Economics and Statistics, 57: 259-268, 1975.

FEA/US Department of Agriculture, ERS. Energy and U.S. Agriculture: 1974 Data Base. Volume 1. 1976.

FEA/US Department of Agriculture, ERS. Energy and U.S. Agriculture: 1974 Data Base. Volume 2. 1977.

Henderson, James M. and Richard E. Quandt. Microeconomic Theory. Third Edition. New York: McGraw-Hill Book Company, 1980.

Hogan, W.W. "Capital-Energy Complementarity in Aggregate Energy-Economic Analysis." Energy Modeling Forum Paper, EMF1. 10, Stanford University, 1978

Hogan, W.W. and A.S. Manne. "Energy-Economic Interactions: The Fable of the Elephant and the Rabbit." In Modeling Energy-Economy Interactions: Five Approaches (ed.) C.J. Hitch, Washington, DC: Resources for the Future, 1977.

Khaleghioseinabadi, G.H. "An Economic Analysis of Energy Demand and Factor Substitution in Western U.S. Agriculture." Ph.D. thesis, University of Hawaii, Honolulu, 1985.

Kliebenstein, James B. and Jean-Paul Chavas. "Adjustments of Midwest Grain Farm Businesses in Response to

Increases in Petroleum Energy Prices." Southern J. of Agr. Econ., 9: 143-148, 1977.

La Bell, P.G. Energy Economics and Technology. Baltimore: Johns Hopkins University Press, 1982.

Landsberg, Hans H. Energy: The Next Twenty Years. Cambridge, MA: Ballinger Publishing Company, 1979.

Miranowski, John A. "Effects of Energy Price Rises, Energy Constraints, and Energy Minimization on Crop and Livestock Production Activities." North Central J. of Agr. Econ.. 1: 1-14, 1979.

Miranowski, John A. and Edward K. Mensah. "Derived Demand for Energy in Agriculture: Effects of Price, Substitution and Technology." Contributed Paper, American Agricultural Economics Association Meetings, 12 pp., 1979.

Pimentel, D. and M. Burgess. "Energy Inputs in Corn Production." In Handbook of Energy Utilization in Agriculture (ed.) D. Pimentel. Boca Raton, Florida: CRC Press, 1980.

Stout, B.A. Energy Use and Management in Agriculture. North Scituate, Mass.: Breton Publishers, 1984.

U.S. Department of Agriculture, ESCS. Energy and U.S. Agriculture: 1974 and 1978. Statistical Bulletin No. 632, 1980.

Contributors

DANIEL J. BERNARDO, formerly Research Assistant, Department of Agricultural Economics, Washington State University, Pullman, WA.

WILLIAM J. CHANCELLOR, Professor of Soil Mechanics, Department of Agricultural Engineering, University of California, Davis, CA.

GLENN S. COLLINS, Formerly Assistant Professor, Department of Agricultural Economics, Texas A&M University, College Station, TX.

RICHARD H. CUENCA, Associate Professor of Soil and Irrigated Crop Production Functions and Hydrology, Department of Agricultural Engineering, Oregon State University, Corvallis, OR.

CHENNAT GOPALAKRISHNAN, Professor of Resource and Marine Economics, Department of Agricultural Economics, University of Hawaii, Honolulu, HA.

RONALD C. GRIFFIN, Assistant Professor of Resource Economics, Department of Agricultural Economics, Texas A&M University, College Station, TX.

ROBERT M. HAGAN, Professor of Irrigation, Water, Soil and Plant Relations, Department of Land Air and Water Resources, University of California, Davis, CA.

JOEL R. HAMILTON, Professor of Production Economics, Department of Agricultural Economics, University of Idaho, Moscow, ID.

WYATTE L. HARMAN, Associate Professor of Farm Management, Department of Agricultural Economics, Texas A&M University, Amarillo, TX.

WARREN E. JOHNSTON, Professor of Natural Resources and Environmental Economics, Department of Agricultural Economics, University of California, Davis, CA.

RONALD D. LACEWELL, Professor of Resource Economics, Department of Agricultural Economics, Texas A&M University, College Station, TX.

DENNIS L. LARSON, Associate Professor of Systems Analysis, Department of Soils, Water, and Engineering, University of Arizona, Tucson, AZ.

WAYNE A. LEPORI, Professor of Energy and Production Mechanics, Department of Agricultural Economics, Texas A&M University, College Station, TX.

BRIAN L. MCNEAL, Professor of Soil Chemistry and Chairman, Department of Soil Science, University of Florida, Gainesville, FL.

VINCENT F. OBERSINNER, Formerly Research Assistant, Department of Agricultural Economics, Washington State University, Pullman, WA.

EDWIN B. ROBERTS, Staff Research Associate, Department of Land, Air and Water Resources, University of California, Davis, CA.

GORDON R. SLOGGETT, Formerly Agricultural Economist, ERS, USDA, Department of Agricultural Economics, Oklahoma State University, Stillwater, OK.

JAMES C. WADE, Associate Professor of Production and Resource Economics, Department of Agricultural Economics, University of Arizona, Tucson, AZ.

GERALD M. WARD, Professor of Ruminant Nutrition, Department of Animal Science, Colorado State University, Fort Collins, CO.

NORMAN K. WHITTLESEY, Professor of Production and Resource Economics, Department of Agricultural Economics, Washington State University, Pullman, WA.

JOHN F. YANAGIDA, Formerly Assistant Professor of Resource Economics, Department of Agricultural Economics, University of Nevada, Reno, NV.

Index